PRAISE FOR SNAKE OIL

Those who think fracked gas is a panacea for our energy future would do well to read this cautionary account—it has an undeniable whiff of reality about it.

— Bill McKibben, founder of 350.org and author of
Oil and Honey: The Education of an Unlikely Activist

Many long-time observers of the world energy scene have been wondering whether claims being made for US shale gas and tight oil are "too good to be true." Here is hard evidence that they are indeed. America will achieve real, long-term energy independence and security only by doing two things: reducing energy demand and developing distributed renewable energy sources.

— Michael Klare, Director of Five College Program in
Peace and World Security Studies at Hampshire
College, author of *The Race for What's Left*

Snake Oil exposes the unsustainable economics behind the so-called fracking boom, giving the lie to industry claims that natural gas will bring great economic benefits and long-term energy security to the United States. In clear, hard-hitting language, Heinberg reveals that communities where fracking has taken place are actually being hurt economically. For those who want to know the truth about why natural gas is a gangplank, not a bridge, *Snake Oil* is a must-read.

— Michael Brune, Executive Director of the
Sierra Club and author of *Coming Clean*

Unconventional production from shales has been hyped mercilessly by the oil and gas industry. Richard Heinberg does an outstanding job of purging the myths and bringing sensibility to a dialogue which, unfortunately, has been driven by a brand of thinking on the part of energy producers that closely mimics the mentality of Wall Street.

— Deborah Rogers, founder of Energy Policy Forum,
former Wall Street financial analyst

We are already living with the false promise that fracking would not harm the environment. Now read the facts about the false promise that it will provide an energy-secure future and lots of jobs. *Snake Oil* debunks all the myths. It is a must-read for our elected leaders.

— Maude Barlow, Board Chair of Food & Water Watch
and author of *Blue Covenant*

It always sounded too good to be true . . . Richard Heinberg authoritatively explains the fine print beneath the hype—sure enough, fracking is the 21st-century version of snake oil. The real deal is energy efficiency along with solar and wind power. It's time for a sober and honest debate about our energy future. That's why *Snake Oil* is essential reading.

— David W. Orr, Paul Sears Distinguished Professor of
Environmental Studies and Politics at Oberlin College,
author of *Down to the Wire*

In *Snake Oil*, Richard Heinberg reveals two key themes everyone should understand: The promise of shale oil and gas abundance is exaggerated and misleading, and the cost of producing the shale oil supply has reached a level that is inconsistent with economic growth.

— Arthur Berman, consulting geologist and Director of
Labyrinth Consulting Services, Inc.

The shale "revolution" might be over sooner than you think. *Snake Oil* does what the mainstream media have failed to do: ask the tough questions—particularly about the economics—and closely examine the real potential of fracking for oil and gas. This book is an important counterpoint to what has been little more than regurgitated industry press releases thus far and is an essential read for investors and policy makers.

— Chris Nelder, energy analyst and author of *Profit from the Peak*

When some things are too hard or painful to understand, it takes a heroic effort to even try to get a message across. Richard Heinberg is a beacon of light cutting through the fog of wishful thinking to illuminate the critical issues by which the human project may stand or fall: energy and economy.

— James Howard Kunstler, author of *Too Much Magic*
and *The Long Emergency*

SNAKE OIL

HOW FRACKING'S FALSE PROMISE OF PLENTY IMPERILS OUR FUTURE

RICHARD HEINBERG

 post carbon **institute**

Cover design by Luke Massman-Johnson
Cover photo of bottle copyright © iStockphoto / Lebazele
Snake head illustration copyright © Moustyk / Dreamstime.com

Printed in the United States of America
First printing July 2013
10 9 8 7 6 5 4 3 2 1

ISBN-13: 978-0976751090
ISBN-10: 0976751097

Post Carbon Institute
613 Fourth Street, Suite 208
Santa Rosa, CA 95404
(707) 823-8700
www.postcarbon.org

This book is dedicated to those activists working to create an energy system where people, places, and the future matter.

CONTENTS

ACKNOWLEDGMENTS

This book relies heavily on the research of many scientists, analysts, and activists. I owe a debt of gratitude first and foremost to my colleague David Hughes, upon whose research much of this book is based, along with Deborah Rogers and Art Berman.

The quick development of such a time-sensitive book was made possible by a community engagement model of publishing. Thanks to the many generous individuals who participated as benefactors and real-time editors. Special thanks to our "Snake Charmers": Ed Adamthwaite, Ruben Bakker, Diane Blust, Brian Bucktin, Clinton Callahan of nextculture.org, Clare Conry, Leonard Edmondson, David J. Fleming, Greg Fox, Felipe Garcia, Peter Gendel, Christopher Gerwin, John B. Howe, Leo Immonen, Kelly Kellogg, John Kretsinger, Richard Larson, Gary Marshall, Chris May, Stephen Miller, Dan Miner, Charles W. Nuckolls, John Parry, Markus Schellenberg, Dennis Schulinck, Richard Seymour, Patti Michelle Sheaffer, Edgar Shepherd, Gary H. Stroy, Nathan Surendran, Jeffrey Tomasi, Richard Turcotte, David Watters, and Matt Wilson. Thanks also to the over three hundred sponsors who joined our "Merry Band of Editors" and provided critical feedback on early drafts; their names are listed at the back of this book.

Many thanks to our lightning-fast production team: Girl Friday Productions, cover artist Luke Massman-Johnson, map designer John Van Hoesen, and research assistant Chris Takahashi. And words of appreciation are in order for Tod Brilliant, for having the initial idea for a new and unique publication model and making it happen; Daniel Lerch, for shepherding the book through our independent publishing process; and Asher Miller, for his editorial suggestions and overall project coordination.

Finally, sincere thanks as always to PJ and LH for their constant and generous support.

Introduction

A FRONT-ROW SEAT AT THE PEAK OIL GAMES

For the past decade I've been a participant in a high-stakes energy policy debate—writing books, giving lectures, and appearing on radio and television to point out how downright dumb it is for America to continue relying on fossil fuels. Oil, coal, and natural gas are finite and depleting, and burning them changes Earth's climate and compromises our future, so you might think that curtailing their use would be simple common sense. But there are major players in the debate who want to keep us burning more.

In the past two or three years this debate has reached a significant turning point, and that's what this book is about. Evidence that climate change is real and caused by human activity has become irrefutable, and serious climate impacts (such as the melting of the Arctic ice cap) have begun appearing sooner, and with greater severity, than had been forecast. Yet at the same time, the notion that fossil fuels are supply-constrained has gone from being generally dismissed, to being partially accepted, to being *vociferously* dismissed. The increasingly dire climate story has achieved widespread (though still insufficient) coverage, but the puzzling reversals of public perception regarding fossil fuel scarcity or abundance have received little analysis outside the specialist literature. Yet, as I will argue, claims of abundance are being used by the fossil fuel industry to change the public conversation about energy and climate, especially in the United States, from one of, "How shall we reduce our carbon emissions?" to "How shall we spend our newfound energy wealth?"

I will argue that this is an insidious and misleading tactic, and that the abundance argument is based not so much on solid data (though oil and gas production figures have indeed surged in the United States), as on exaggerations about future production potential, and on a pattern of denial regarding steep costs to the environment and human health.

The change in our public conversation about energy is predicated on new drilling technology and its ability to access previously off-limits supplies of crude oil and natural gas. In the chapters ahead, we will explore this technology—its history, its impacts, and its potential to deliver on the promises being made about it. As we will see, horizontal drilling and hydrofracturing ("fracking") for oil and gas pose a danger not just to local water and air quality, but also to sound energy policy, and therefore to our collective ability to avert the greatest human-made economic and environmental catastrophe in history.

★ ★ ★

Permit me to use a metaphor to further frame the discussion we'll be having about fossil fuel abundance or scarcity. Since all debates are contests, at least superficially, it's possible to summarize this one as if it were a game—like a soccer match or a bowling tournament. Of course, it is far more than just a game; the stakes, after all, may amount to the survival or failure of industrial civilization. But games are fun, and it's easy to keep track of the score. So . . . let the metaphor begin!

First, who are the teams? On one side we have the oil and gas industry, its public relations minions and its bankers, as well as a few official agencies—including the US Energy Information Administration (EIA) and the International Energy Agency (IEA)—that tend to parrot industry statistics and forecasts. This team is respected and well funded. For reasons that will become apparent in a moment, we'll call this team "the Cornucopians" (after the mythical horn of plenty, an endless source of good things).

The other team consists of an informal association of retired and independent petroleum geologists and energy analysts. This team has little funding, is poorly organized, and hardly even existed as a recognizable entity a decade ago. This is my team; let's call us "the Peakists" (in reference to the observation that rates of extraction of nonrenewable resources tend to peak and then decline).

These two teams have very different views of the energy world. Back in 2003, the Cornucopians were saying that global oil production would continue to increase in the years and decades ahead to meet rising demand, which would in turn grow at historic rates of about 3% per year (about the same rate at which the economy was expanding). Meanwhile oil prices would stay at approximately their then-current level of $20 to $25 per barrel.[1] The Cornucopians' message could be summarized as "There's nothing to worry about, folks. Just keep driving."

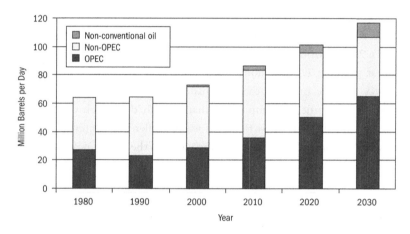

Figure 1. World Oil Production Forecast to 2030 (Cornucopians).

Source: International Energy Agency, World Energy Outlook 2003.

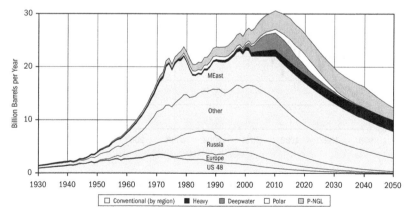

Figure 2. World Oil Production Forecast to 2050 (Peakists).

Source: Colin Campbell, Association for the Study of Peak Oil and Gas, July 2003.

This view was in stark contrast to that of us Peakists, who, based on geological evidence from around the world (depleting older supergiant oil fields, declining rates of discovery of new fields, and increasing costs to develop them), were saying that rates of global oil production would soon reach a maximum and start to diminish, while petroleum prices would soar.[2] The Peakists' argument wasn't that the world would suddenly *run out* of oil anytime soon, but that the end of *cheap* oil and *expanding rates of production* was approaching. Since oil price spikes have had severe economic impacts in recent decades, the implication was clearly that societies would be better off weaning themselves from oil as quickly as possible.

Well, what has actually happened? How has the game progressed so far?

In 2005, world crude oil extraction rates effectively stopped growing. In that year the average global production rate was 73.8 million barrels per day (mb/d); in 2012, that rate had only increased to 75.0 mb/d—a relatively insignificant bump of less than 1.5 mb/d in seven years (a 0.3% average annual rate of growth). This was completely counter to the forecasts of the Cornucopians, but it fit the views of the supply pessimists well. Point for the Peakists.

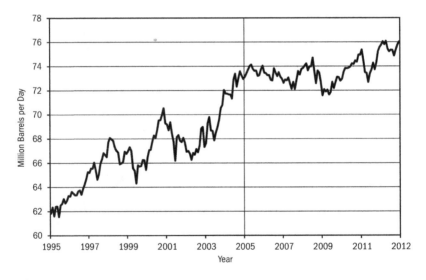

Figure 3. World Crude Oil Production, 1995–2012. World oil production growth tapered off markedly after 2005.

Source: Energy Information Administration, 2013. Data include lease condensates and exclude natural gas plant liquids, refinery process gain, and biofuels.

With oil supply rates stagnant, prices went up—soaring from a yearly (inflation adjusted) average of $35 per barrel in 2003 to a yearly average of $110 in 2012. Again, this development was completely unforeseen by Cornucopians but had been clearly and repeatedly forecast by Peakists. Point for my side.

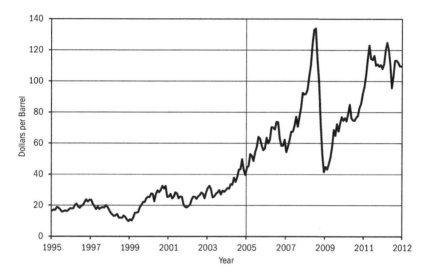

Figure 4. Brent Crude Oil Price, 1995–2012. Oil prices started surging past historic highs just prior to 2005.
Source: http://www.indexmundi.com.

When the world oil price briefly shot up to nearly $150 per barrel in the summer of 2008, the global economy shuddered and swooned. Thus began the worst recession since the 1930s. Of course, other factors contributed to the crash—most notably, a bursting housing bubble in the United States and an unsustainable buildup of debt in nearly all the world's industrial economies. But it's clear both that high oil prices added to financial instability, and that the oil price spike of 2008 provided a sudden gust that helped bring down the house of cards. Peakists had been warning of the economy's vulnerability to high oil prices for years; here was dramatic confirmation. Another point for my team.

Now we've arrived at the period 2008–2009; at that stage of the game, the score was Peakists 3, Cornucopians zip. Despite the fact that we Peakists had virtually no funding and limited media access, we were

seriously in danger of winning the debate. The term *peak oil* went from
being unknown, to being associated with conspiracy theorists, to being
broadly familiar to those who followed energy issues.

The Cornucopians, however, were not about to throw in the towel.
In fact, they were just shaking off the complacency that accompanied their
status as reigning champs. And they were about to deploy a significant
new game strategy.

The "peak" issue was not limited to oil. US conventional natural
gas production had been declining for years, and prices were soaring.
Peakists said this was evidence of an approaching natural gas supply crisis.[3] Instead, high prices provided an incentive for drillers to
refine and deploy costly hydraulic fracturing technology (commonly
referred to as "fracking") to extract gas trapped in otherwise forbid-
ding shale reservoirs. Small- to medium-sized companies crowded
into shale gas plays in Texas, Louisiana, Arkansas, and Pennsylvania,
borrowed money, bought leases, and drilled tens of thousands of wells
in short order. The result was an enormous plume of new natural gas
production. As US gas supplies ballooned, TV talking heads (reading
scripts provided by the industry) and politicians all began crowing
over America's "game changing" new prospect of "a hundred years
of natural gas." We Peakists hadn't foreseen any of this. Point to the
Cornucopians.

Not only did supplies of natural gas grow, but prices plummeted. In
the pre-fracking years of 2001 to 2006, gas prices had shot up from their
1990s level of $2 per million Btu to over $12 (Figure 6). But after 2007,
as the hydrofracturing boom saturated gas markets, prices plummeted
back to a low of $1.82 in April 2012. Gas was suddenly so cheap that
utilities found it economic to use in place of coal for generating base-
load electricity. The natural gas industry began to promote the ideas of
exporting gas (even though the United States remained a net natural
gas importer), and of using natural gas to power cars and trucks. Again,
Peakists had completely failed to forecast these developments. Point
Cornucopians.

Then, using the same hydrofracturing technology, the industry
began to go after deposits of oil in tight (low-porosity) rocks. In Texas
and North Dakota, US oil production began growing. It was an aston-
ishing achievement, especially since the nation's oil production had

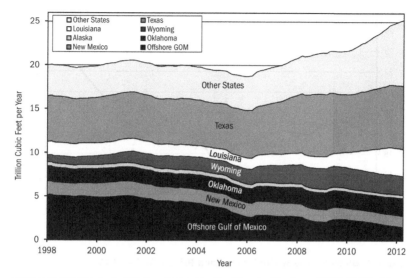

Figure 5. US Marketed Natural Gas Production by Region, 1998–2012. Oil prices started surging past historic highs just prior to 2005.

Source: J. David Hughes, "Drill, Baby, Drill," Figure 18; data from Energy Information Administration, December 2012, fitted with 12-month centered moving average. Note that marketed production is wet gas and includes gas used for pipeline distribution and at gas plants and leases that is not available to end consumers.

generally been declining since 1970 (Figure 7). Suddenly there was serious discussion in energy policy circles of America soon producing more oil than Saudi Arabia. None of us Peakists had predicted this. Point Cornucopians.

That brings us to the present. As of 2013, the game is tied and headed into overtime. Cornucopians have the momentum and the historic advantage, so they've been quick to claim victory. Meanwhile, at least one prominent Peakist has publicly conceded defeat: in a widely circulated essay, British environmental writer George Monbiot recently proclaimed that "We were wrong on peak oil."[4]

It doesn't look good for my team. It appears to most people that the "Shale Revolution" (the tapping of shale gas and tight oil, thanks to advanced drilling techniques) has changed the game for good. Is it time for us to exit the playing field, heads bowed, shoulders slumped?

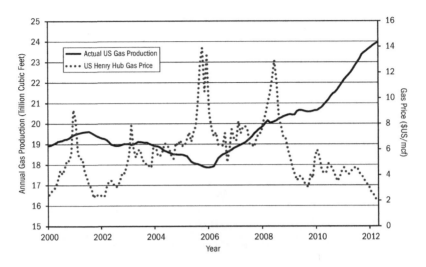

Figure 6. US Natural Gas Production and Prices, 2000–2012.

Source: Adapted from J. David Hughes, "Drill, Baby, Drill," Figure 34; data from Energy Information Administration, December 2012. Production data fitted with 12-month centered moving average.

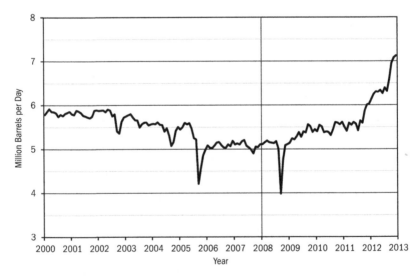

Figure 7. US Crude Oil Production, 2000–2013. US oil production reversed decades of decline in 2008 and then surged in late 2011.

Source: Energy Information Administration, May 2013. Data include lease condensates and exclude natural gas plant liquids, refinery process gain, and biofuels.

★ ★ ★

As you've probably guessed from the title of this book, the pages that follow are not intended as a capitulation. Rather, my purpose is to alert readers to relevant and important information that is, with rare exceptions, failing to find its way into the public discussion about our energy future. Its upshot is that *the game is about to turn again.*

Almost no one who seriously thinks about the issue doubts that the Peakists will win in the end, no matter how pathetic my team's prospects may look for the moment. After all, fossil fuels are finite, so depletion and declining production are inevitable. The debate has always been about timing: Is depletion something we should worry about now?

Readers who've seen articles and TV ads proclaiming America's newfound oil and gas abundance may find it strange and surprising to learn that the official forecast from the US Energy Information Administration (EIA) is for America's historic oil production decline to resume *within this decade.*[5]

But the EIA may actually be overly optimistic. Once the peak is passed, the agency foresees a long, slow slide in production from tight oil deposits (likewise from shale gas wells). However, analysis that takes into account the remaining number of possible drilling sites, as well as the high production decline rates in typical tight oil and shale gas wells, yields a different forecast: production will indeed peak before 2020, but then it will likely fall much more rapidly than either the industry or the official agencies forecast.

There's more—much more. This book tells an analytic story assembled from proprietary industry data on every active and potential US oil and gas play. It's a story about shale gas wells that cost more to drill than their gas is worth at current prices; a story about Wall Street investment banks driving independent oil and gas companies to produce uneconomic resources just so brokers can collect fees; and a story about official agencies that have overestimated oil production and underestimated prices consistently for the past decade.

The book also relates a human and environmental story gathered from people who live close to the nation's thousands of fracked oil and gas wells—a tale of spiraling impacts to drinking water, air, soil, livestock, and wildlife; about companies failing to pay agreed lease fees; about declining property values; about neighbor turned against neighbor; and about boom towns in turmoil.

Here, in briefest outline, are the findings the evidence supports:

- The oil and gas industry's recent unexpected successes will prove to be short-lived.
- Their actual, long-term significance has been overstated.
- New unconventional sources of oil and gas production come with hidden costs (both monetary and environmental) that society cannot bear.

Further, these conclusions lead inevitably to one final, crucial observation:

- The oil and gas industry's exaggerations of future supply have been motivated by short-term financial self-interest, and, to the extent that they influence national energy policy, they are a disaster for America and for future generations.

★ ★ ★

This book is aimed at the general public and at policy makers, who need to understand why the current received wisdom about US fossil fuel abundance is dangerously wrong.

It is especially directed toward local anti-fracking activists across the United States and throughout the world who are working hard to limit or prevent harms to water and air quality, wildlife, and human health. Bolstering environmental arguments with economic data showing the likely brevity of the fracking boom can only help win debates regarding the regulation of this dangerous technology.

The book is meant as well for the thousands of readers who learned about peak oil during the past decade, took the information seriously, and made extraordinary efforts to reduce their personal petroleum dependency and to prepare their communities for the end of the era of cheap oil—only to see their credibility erode as a result of oil and gas industry disinformation and spin. These are my people, and they need some encouragement right about now.

Finally, and perhaps most importantly, this book is directed toward anyone and everyone who cares about the fate of our planet. The only realistic way to avert catastrophic climate change is to dramatically and quickly reduce our consumption of fossil fuels. That project will pose economic and technical challenges. But politics may present the biggest obstacle of all.

There are two kinds of arguments for policies to reduce reliance on oil, coal, and gas—environmental and economic. *Environmental* arguments point to the consequences of rising greenhouse gas emissions from burning hydrocarbons, including rising sea levels, extreme weather, and likely catastrophic impacts to agriculture. *Economic* arguments highlight the inevitability of future fossil fuel scarcity as society burns these finite, nonrenewable resources in ever-greater quantities. The clear solutions in both cases: find other energy sources and reduce overall energy consumption *now*.

The fossil fuel industry has, quite understandably, fought back against both economic and the environmental arguments. Oil companies (notably ExxonMobil) have not only funded the efforts of climate-denial front groups to sow doubt about what is in fact established science (ExxonMobil now officially acknowledges the reality of human-induced climate change), they have also mounted a sustained public relations campaign to undermine the credibility of peak oil analysts. At the same time, the industry would like nothing better than to divide its opponents, and it has achieved some success in this regard: a few climate activists have mistakenly disavowed peak oil, perhaps because they see it as a distraction from, or dilution of, their own message. They often point out that if industry estimates of fossil fuel reserves are correct, burning all that oil, coal, and gas will result in environmental destruction on a scale beyond our ability to comprehend; with so much at stake, why quibble about when oil production rates will max out? Meanwhile, a few Peakists have made the foolish claim that climate change is not a serious problem because the global economy will crash due to soaring energy prices before we are able to do really serious damage to the environment.

Success in shifting energy policy depends upon *coordination of* environmental and economic arguments against continued reliance on fossil fuels. Are there enough accessible hydrocarbons to tip the world into climate chaos? Absolutely. But activists concerned about climate change would do well to embrace economic (supply constraint) arguments against fossil fuel dependency. By erroneously reinforcing industry hype about the future potential of shale gas, tight oil, and tar sands, they keep the debate exactly where the industry wants it—as a choice between environmental protection on the one hand and jobs, economic growth, and energy security on the other. It's a false choice and a losing strategy.

★   ★   ★

Here's what readers can expect to find in the pages ahead. After a quick overview in Chapter 1 of what the peak oil and gas discussion is all about and why it matters, we will take a close, hard look in Chapter 2 at fracking—what it is and what it means. In Chapter 3, we'll examine key producing regions, the rates at which per-well output tends to decline over time, and trends in drilling. And we will explore the implications of those data.

We will then look at the environmental costs of unconventional oil and gas in Chapter 4, sampling reports from the front lines of the fracking fields across the United States regarding impacts to water and air quality, land, and public health. You may be surprised to learn who is fighting the drilling juggernaut—it's not just environmentalists.

In Chapter 5, we'll inquire who actually benefits from the fracking boom and explore Wall Street's role in the current mania. Investment bankers make money on the way up (as bubbles inflate) and on the way down (as companies sell off assets and submit to mergers and takeovers). Therefore, it is in their interest to support drillers' exaggerated claims for reserves and future production potential. When the fracking boom inevitably goes bust, it won't be the banks that will take the hit; it will be the investors (including retirees) who bought shares of stock in oil and gas companies.

Finally, in Chapter 6, we will examine other unconventional fuels and fuel sources (tar sands, methane hydrates, and oil shale) to see whether they might be game changers waiting in the wings. And we will explore likely scenarios for our *real* energy future. (Just one preliminary hint: it's time to learn how to live well with less.)

★   ★   ★

The data we will survey in the chapters ahead suggest that, through the technology of hydrofracturing, the oil and gas industry will generate *10 or fewer years of growing fuel supplies.* (In the case of shale gas, the clock started ticking roughly five years ago; for tight oil, about three years ago). Industry promises of a hundred years of cheap, abundant gas, and of domestic oil production growth making the nation self-sufficient in petroleum are unlikely to be fulfilled given what we know now about the nature of the resources and the technology being used to access them.

Let me be clear: I am not saying that the United States will *run out* of shale gas or tight oil sometime in the next five to seven years, but that the current spate of oil and gas *supply growth* will probably be over, finished, done and dusted before the end of this decade. Production will start to decline, perhaps sharply.

Meanwhile the brief, giddy production boom we are currently seeing in towns, farms, and public lands in Texas, North Dakota, Pennsylvania, and a few other states will have come at an enormous cost. In order to achieve just a few years of domestic supply growth, the industry will need to drill tens of thousands of new wells (in addition to the tens of thousands brought on line in just the last three to five years), ruining landscapes, poisoning water, and forcing families to abandon their homes and farms.

This temporary surge of production may yield a very few years of lower natural gas prices and may temporarily improve the US balance of trade by reducing oil imports. What will we do with those years of reprieve? In the best instance, the fracking that has already been accomplished could provide us a bonus inning in which to *prepare for life without cheap fossil energy.* But to make use of this borrowed time we must build an energy infrastructure of wind turbines and solar panels rather than drilling rigs and pipelines. This will constitute the biggest investment, and the most ambitious project, of our lifetimes. Currently, instead, many renewable energy efforts are being hampered by the false perception of vast, long-term supplies of cheap natural gas.

We are starting the energy transition project of the 21st century far too late to altogether avert either devastating climate impacts or serious energy supply problems, but the alternative—continued reliance on fossil fuels—will ensure a future far worse, one in which even the bare survival of civilization may be in question. As we build our needed renewable energy system, we will also need to build a new kind of economy, and we must make our communities far more resilient, so as to withstand environmental and economic shocks that are inevitably on their way.

Meanwhile the fossil fuel industry is doing everything it can to convince us we don't have to do anything at all—other than simply to keep on driving. The purveyors of oil and natural gas are selling products that we all currently use and that we still depend upon for our modern way of life. But they're also selling a vision of the future—a vision as phony as the snake oil hawked by carnival hucksters a century ago.

SNAKE BITES

 THE INDUSTRY SHILLS SAY:
Peak oil is crap. *World oil reserves are increasing.*

THE REALITY IS:
The industry has *overstated* world oil reserves by about a third and is working harder and harder just to stand still.

② THE CONVENTIONAL WISDOM SAYS:
Unconventional oil (tar sands, tight oil) will seamlessly replace the current energy output from conventional sources.

THE REALITY IS:
It takes energy to get energy. The energy return on energy invested (EROEI) of unconventional fossil fuels is significantly worse than for conventional resources.

*Oil production technology is giving us **ever-more expensive oil** with **ever-diminishing returns** for the **ever-increasing effort** that needs to be invested.*

~ Raymond Pierrehumbert, Professor of Geophysical Sciences, University of Chicago

Chapter One

THIS IS WHAT PEAK
OIL LOOKS LIKE

Oil is the linchpin of our modern industrial way of life. Nearly all energy used for transport derives from it, and transport is essential to virtually all trade. Take petroleum away and the global economy would shudder to a halt in a matter of minutes.

It wasn't always this way. The petroleum age started when the first commercial oil well was drilled in the late 1850s, and it wasn't until the early 20th century that energy-dense, easily portable "rock oil" found widespread use.

With automobiles, airplanes, tractors, chainsaws, diesel-fueled trains, oil-powered ships, and diesel-powered mining and road-building equipment, it became possible to intensify and expand nearly every extractive and productive process known to humankind—including the process of drilling for oil. Agriculture, fishing, mining, transportation, manufacturing, and trade burgeoned as never before, lifting billions from poverty (or undermining their more sustainable traditional ways of life, depending on how you look at it) and providing several hundred million humans with a level of amenity, convenience, and mobility undreamt of even by the pharaohs and emperors of previous eras.

All these benefits have come at a cost. The growth of extractive industries has led to increasing rates of depletion of minerals, soil, water, fish, and forests. At the same time, the expansion of industry has created burgeoning streams of waste products that nature cannot absorb. The most pervasive of industrial wastes is carbon dioxide, released when oil and other fossil fuels are burned. As ambient levels of carbon dioxide rise, the planet's atmosphere traps more heat, changing the climate and precipitating extreme weather events, potentially leading to conditions in which civilization cannot persist.

Resource depletion and climate change are problems that undermine the survival prospects of future generations. Many people still tend to think their impacts are decades away, but we are beginning to see those impacts unfold in real time all around us.

In this chapter, we will take a whirlwind tour of a controversy that has roiled the oil and gas industry for the past decade and more. The discussion about oil supplies, reserves, and production that goes by the name of "peak oil" is complex and subtle and has often been mischaracterized or ludicrously oversimplified ("We're running out!" versus "We're not running out!"). It is a discussion that readers must be familiar with in order to properly understand and evaluate the claims of abundance currently being made by representatives of the fossil fuel industry. What follows is intended both as an overview of the current state of that discussion, and as an effort to set the record straight with regard to economic life-or-death issues that have sometimes been distorted by those who profit from our fossil-fueled status quo.

PEAK OIL: WHAT THE FUSS IS ABOUT

Individual oil wells have a finite life span. Sometimes aggressive techniques—such as water flooding—can be deployed to extend an oil well's life, but depletion is inexorable, and eventually every oil well reaches the point where production rates decline severely and the cost of extraction efforts exceeds the value of the oil being extracted. When that happens, the well is capped or plugged with cement, and equipment is removed from the site.

The same principle holds for larger aggregations of petroleum resources. When a new oil field is discovered, a few exploratory wells are drilled to help determine the size of the deposit and the nature of the geology. With this information, engineers determine optimum well placement and start drilling in earnest. The production rate for the oil field increases as more wells are drilled. Gradually, as older wells deplete, their production rates begin to decline, but new wells are drilled to offset those declines. Eventually, when all of the possible drilling locations have been used and production from most wells is tailing off, it becomes impossible to stave off the dwindling of the overall extraction rate.

Bundle many oil fields together and again the same principle holds. At this level of scale a pattern becomes apparent: the aggregate oil extraction rate begins to approximate a bell curve. The top of the curve represents the maximum production rate, or the peak of oil production, for the fields in question.

Most oil-producing countries—including Indonesia, the United Kingdom, Norway, and the United States—saw their national peaks in crude oil production years or decades ago. Their production declines have been offset by discoveries and production growth elsewhere in the world.

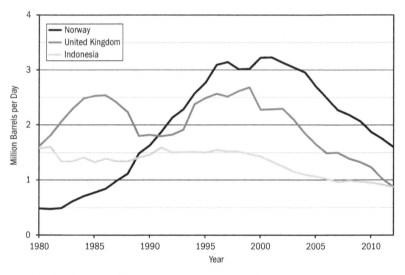

Figure 8. Norway, UK, and Indonesia Oil Production, 1980–2012.

Source: Energy Information Administration, May 2013.

But there are only so many potential oil-producing areas on our small planet. Therefore, the same peaking trend will inevitably hold for the entire world. The rate of global oil production will rise to a plateau or peak, then decline. Unless we have somehow substantially reduced our dependency on oil by the time that decline commences, the impact to the global economy will be serious-to-catastrophic. Therefore the *timing* of peak oil is of great importance.

And that's what all the fuss is about.

OKAY, WHEN?

Oil analysts have two main ways of forecasting the timing of the global peak. One involves applying a fairly simple equation to past and current production statistics; the other is a more detailed method of adding likely flows from potential new sources and subtracting declines from existing fields (which number in the thousands). Neither method is foolproof. The data are too complex to permit the accurate forecasting of the global oil production peak to the day, month, or year. But many analysts agree that around 2005, as global crude oil production hit a plateau that continues to the present, our world entered *the peaking period*, and within a few years the global oil production rate will in all probability start to decline.

Here are some of the factors that complicate efforts to forecast the peak:

Reserves and resources. Some analysts (in the Introduction we called them "Cornucopians") are highly optimistic about oil's future. They typically point to enormous reserves of oil around the world, which continue to grow—for reasons discussed below. If there's all that oil left to extract, they ask, does it make sense to worry about an imminent peak in production rates?

Peakists reply that focusing on reserves numbers can be misleading, as not all oil is the same. Saudi oil, most of which was discovered in the 1950s and 1960s, can be produced cheaply and quickly; the oil being brought on line now from tar sands, deepwater, and tight formations will either be extracted slowly, or will require high levels of investment, or both.

Sometimes reckless oil boosters confuse *reserves* (defined as the portion of the total hydrocarbon endowment that is extractable at realistic market prices and with current technology) with *resources*—the total endowment. For example, the Green River shale formation in Colorado represents a resource base equivalent to roughly a trillion barrels of oil. If all those resources were to be counted as reserves, the United States would instantly leap to the top of the list of oil-bearing nations. However, with current technology and at current market prices, virtually no oil is being commercially produced from the Green River formation, and that situation is not likely to change anytime soon. (We'll see why in Chapter 6.) The point cannot be overemphasized: *the peak oil discussion is about rate of supply, not size of resources or even reserves.*

New discoveries. Cornucopians often trumpet new oil discoveries, such as ultra-deepwater finds off the coast of Brazil, as undermining the notion of near-term global production peak. This would be the case if the *rate* of discovery of new oil sources were increasing (but it is not) and if the amount of new oil being discovered annually exceeded the amount being extracted from known fields annually (it doesn't, by a long shot). The peak rate of discovery, when many large fields were being found each year, occurred globally in the early 1960s. In recent years the oil industry has found (on average) one barrel of new oil for every four or five consumed.

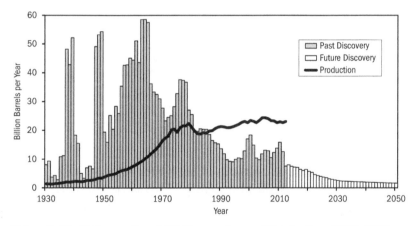

Figure 9. World Conventional Oil Discoveries and Production, 1930–2050.
Source: Colin Campbell, Association for the Study of Peak Oil, 2012.

Cornucopian analysts insist once again that declining discoveries are not a problem because *world oil reserves are increasing.* However, some of this growth in reserves is illusory. In March 2012, Sir David King's team at Oxford University's Smith School of Enterprise and the Environment published a peer-reviewed paper in *Energy Policy*, concluding that the industry had overstated world oil reserves by about a third.[1] Most of the rest of recent reserves growth has come from reclassification of marginal resources. This has happened partly because of refinements in production technology, but mostly because oil prices have risen high enough to justify the enormous investments required to extract and process tar sands, heavy oil, and tight oil. The upshot: as reserves of regular conventional oil are consumed, they are being replaced by reserves of oil or bitumen that will be produced more slowly, at higher cost, with higher environmental risks, and with the requirement for larger investments of energy into the process of production.

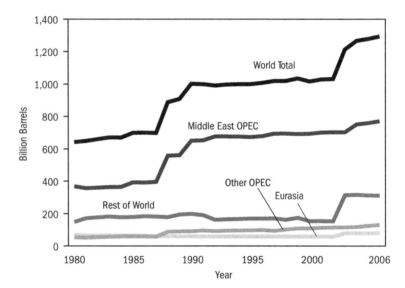

Figure 10. World Declared Oil Reserves, 1980–2006. Reserves include crude oil (including lease condensates) and natural gas plant liquids. Most upward revisions come from OPEC claims of new reserves (whose validity is hotly debated) and reclassification of tar sands and extra-heavy oil deposits as oil reserves. *Source: Energy Information Administration,* International Oil Outlook 2006.

When a new source of supply comes on line it must first replace declines from existing fields before it can help boost overall production to a higher level. For the world as a whole, the rate of decline in production from existing oil fields is between 4 and 5% annually.[2] Thus, every three years the world needs to find brand new sources of oil that, taken together, are as productive as Saudi Arabia just to maintain a constant overall rate of production. For the past seven years, the global oil industry has been able to maintain a rough balance between declines and increases, but doing so has required substantially increased rates of drilling and capital expenditure.

Wild cards. The likely timing of the commencement of the inevitable global oil production decline will be determined not just by geology, but also by wild cards like technology, politics, and the economy.

Technology can make oil accessible that wasn't previously (as is the case of tight oil, which we will discuss in Chapters 2 and 3). On the other hand, political events can take oil production off-line rapidly, as happened in the 1970s with the Arab oil embargo, and as is occurring today with US sanctions against Iranian oil exports. At the same time, the condition of the economy affects oil demand: if the economy booms, demand goes up, and that leads to higher oil prices. Higher prices then stimulate efforts to produce oil that was previously uneconomic. If the economy falters, the price of oil drops and so do efforts to produce marginal sources.

Defining "oil." Another complication arises from the definition of the word *oil*. Does it refer only to conventional crude? When the US Energy Information Administration (EIA) releases statistics for current or future US oil production, the numbers always include *refinery gains*. When oil is refined, the volume of the products yielded from it (as measured in barrels or cubic meters) is greater than the volume of the crude oil that entered the refinery because the refined products are lighter and less dense than crude. Yet refining oil into gasoline, diesel, and kerosene requires energy—thus energy has been *lost* in the process, even though product volume has increased. Moreover, in the EIA's "oil" production statistics for the US, refinery gains for *imported* oil are implicitly lumped together with volumetric gains from the refining of *domestically produced* oil, thus making it seem as though the nation is producing more oil domestically, and importing less, than is really the case.

Further, the International Energy Agency's definition of "oil" includes *natural gas liquids* (or NGLs). Natural gas is mostly methane, but as it comes out of the ground, it may also contain hydrocarbons with longer molecular chains, including propane and butane. These are typically captured in processing and used for heating and for industrial purposes, including the making of plastics. They are called NGLs not because they are liquid at room temperature and normal atmospheric pressure (they're not), but because they can be liquefied at lower pressure and higher temperature than methane and are typically bottled and sold in liquefied, pressurized form. (NGLs are not the same as liquefied natural gas—or LNG—which is methane that has been super-cooled and highly pressurized, usually to make it easier to transport by tanker.) NGLs have only about 60% of the energy by volume as crude oil and are, for the most part, used for purposes different from those of crude oil. So why should NGLs be called "oil"?

Then there are biofuels. These *are* used for purposes similar to those of crude oil—principally, as transportation fuels. However, ethanol and biodiesel are not extracted from the ground; they are made from agricultural products in a process that requires lots of oil and natural gas. Their production uses so much energy, in fact, that it is questionable whether they provide a net energy benefit, and, in any case, counting the biofuels along with the fuels that are consumed in their production is an improper double-counting.

When official agencies call NGLs and biofuels "oil," statistics then show world "oil" production increasing in recent years; when these substances are subtracted from the accounting, nearly all that growth disappears. Thus an important economic signal is often hidden behind statistical noise.

All of this makes it more difficult to answer the question, "When will world oil production begin to decline?" Yet the question loses none of its criticality. Even if forecasting the exact date of the peak is a fool's errand, only a fool would miss the signs that the world oil industry has entered a new, desperate era. Discoveries are down, costs are up. Production has flatlined, environmental impacts from petroleum operations are soaring.

Is this what peak oil looks like?

IT GETS WORSE

The actual peak in world oil production will presumably occur over the course of several years, while the decline in production will continue for decades. Given so much time, one might assume that civilization will gradually adapt without too much stress and strain. Two dilemmas make this much less likely by reducing the time available for adaptation.

The first is the *net export dilemma*. Trade in petroleum is integrated and global. Enormous amounts of oil are shipped by tanker from continent to continent or flow from country to country via pipeline. Many nations (such as Japan) produce no oil and import all they use, while others (like Saudi Arabia) are substantial exporters of petroleum. As the price of oil rises, the revenues to oil exporting nations grow (Saudi Arabia makes more money)—and economic expansion within these nations brings more domestic demand for oil. Thus, exporting nations end up using more of their own oil and exporting less, even if production holds steady from year to year. Indeed, oil demand within Saudi Arabia is growing faster than in all but a few other nations.

Since 2005, as world crude oil production has stayed essentially flat, the amount of petroleum exported has declined by about 5%.[3] Competition for these available exports has nudged oil prices higher. Because industrializing nations like China are able to afford a higher

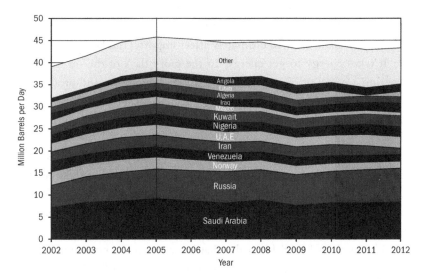

Figure 11. Oil Net Exports of Top 45 Net Exporters, 2002–2012. Net exports have declined over 5% since 2005.

Source: Energy Information Administration, June 2013; compiled by Jeffrey Brown and Daniel Lerch.

price (they haven't spent decades getting used to using *cheap* oil in large quantities), they have effectively outbid older industrialized nations like the United States and most European countries: China imports more, while the US imports less. In the United States and Europe, high oil prices slow the economy, and in a slow economy motorists cut back on driving. Indeed, US drivers have cut back on gasoline consumption. As peak oil blogger Gail Tverberg has noted, American oil consumption in 2012 was about 20% lower than it would have been if the pre-2005 trend in oil consumption growth of 1.5% per year had continued.[4] Some of this reduction is as a result of improved vehicle fuel efficiency, but Americans are also simply driving less.[5]

If this trend toward declining petroleum exports continues, and there is no persuasive reason it won't, then the amount of oil available on the world export market will shrink rapidly over the next decade. Oil importing nations will increasingly be shut out, and older industrialized nations (the United States, Japan, and Europe) will bear the brunt of disappearing export volumes. Petroleum geologist Jeffrey Brown calculates that if current trends were to persist, the United States and Europe would effectively be shut out of the world petroleum export market by 2025.[6]

A second dilemma, *the decline in the energy returned on the energy that's invested in obtaining oil,* will ultimately affect importers and exporters alike.

It takes energy to get energy. In the glory days of the oil industry, investing a barrel of oil's worth of energy in exploration and production yielded a hundred barrels of oil or more over an oil well's lifetime. Today in the US oil patch, the ratio of energy yield to energy investment is closer to 10:1.[7]

Clearly, when the overall energy return on energy invested (EROEI) for the process of oil extraction declines to 1:1, then the oil produced will cease to be an *energy source* in the true sense. It may still be useful as raw material or lubricant, but it will no longer serve to increase the amount of net energy available to do work for society.

The math of EROEI reveals what has come to be known as the *net energy cliff.* At first thought, it might appear that a 100:1 EROEI is 10 times more beneficial to society than a 10:1 energy profit. But it turns out that there's a practical turning point at around 10:1 (the cliff). Above that ratio (from 11:1 to 200:1 and beyond) each incremental increase in EROEI delivers relatively smaller benefit. Below 10:1, each increment of decline is much more decisively detrimental.

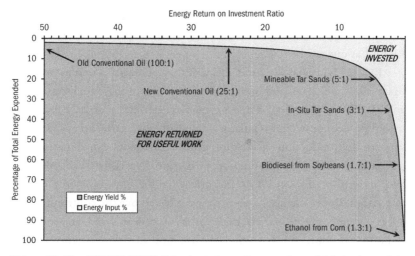

Figure 12. The "EROEI Cliff." This chart shows the energy available to do useful work as a proportion of total energy expended for various resources.
Source: Adapted from J. David Hughes, "Drill, Baby, Drill," Figure 38.

Think of net energy in terms of the number of people in society engaged in energy production. If EROEI = 1:1, then everyone is involved in energy production and there is no one available to take care of society's other needs. If the EROEI is 100:1, then 1 person is involved in energy production and 99 are able to do other things— build houses, teach, take care of the sick, cook, sell real estate, and so on. If we have 2 energy workers and 98 folks doing other things, then EROEI = 50:1; similarly, with 4 folks getting energy and 96 doing other things, EROEI = 25:1. With 8 getting energy and 92 doing other things (EROEI = 12.5:1) there may begin to be problems finding enough folks who are trained at getting energy to provide for all the others, whose every activity uses energy. With 16 getting energy and 84 doing other things (EROEI = 6.25:1) serious problems may become apparent, and an industrial-style organization of society may be only marginally viable.

Agriculture, education, health care, defense, entertainment, transportation, and manufacturing are all *users* of energy. A modern industrial nation needs a big surplus of energy from its energy-production efforts in order to power all these enterprises. White's Law, arguably as important in the field of human ecology as the laws of thermodynamics are in physics, states that the level of economic development possible in any society is determined by the amount of *net* energy available per capita.[8] We ignore EROEI at our peril.

EROEI is crucial to considerations of the potential economic benefits of tar sands, oil shale, and biofuels, because each of these fuel sources has an EROEI of 5:1 or less. Their production can be financially profitable in certain economic and regulatory environments: government subsidies support the production of ethanol, gullible investors can sometimes be persuaded to fund the production of marginal resources like oil shale, and high oil prices can create incentives for the expansion of tar sands operations. But by themselves—if we were to remove the contributions of energy from conventional oil, gas, coal, hydropower, wind, and solar and ramp up tar sands, oil shale, and biofuels instead—these energy sources would be unable to power a complex society.

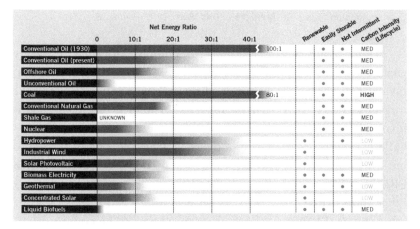

Figure 13. Characteristics of Energy Resources.

Source: Data compiled by David Murphy, from Tom Butler and George Wuerthner, eds., ENERGY: Overdevelopment and the Delusion of Endless Growth, (Healdsburg, CA: Watershed Media, 2012).

Since the EROEI of oil is declining rapidly due both to the depletion of easy-to-produce deposits and to the increasing use of tar sands and tight oil, it would seem sensible for energy policy makers to promote an equally rapid transition to wind and other energy sources that have higher energy returns. However, in reality, renewable energy sources are making only small and slow inroads except in a very few countries (such as Germany and Denmark). That's partly because of the political influence of the fossil fuel industry, but it also results from our enormous sunken investments in the current energy infrastructure of highways, internal combustion engines, gas furnaces and stoves, natural-gas-burning power plants, and so on. Replacing these takes time and money. These sunken investments ensure we will probably continue using oil and gas for decades, despite their deteriorating economics.

Can technology solve the EROEI problem? New ways of extracting oil and gas could increase the energy efficiency of the process. For example, pad drilling and increased rig mobility have enabled drillers in the Eagle Ford formation to reduce average drilling time from 23 days in 2011 to 19 days in June 2012. Reduced drilling time almost certainly translates to reduced energy investment. However, efficiency efforts must push against the tide of declining resource quality: overall, tight oil plays require more energy investment in drilling than conventional oil plays, and as the best resources in those plays are drilled first, each new well requires more effort per unit of productivity. (We will return to this topic in more detail in Chapter 6.)

The relentless decline in EROEI of oil is one of the biggest underreported economic stories of our times. Available net energy—what makes society work—is dwindling away even as production statistics *seem to show* a North American oil and gas production boom.

Add net exports and net energy together, and the situation, especially for industrialized oil importing nations, starts to look pretty severe even over the short term (from now to 2020).

MAKING SENSE OF HISTORY AS WE LIVE IT

From the start of commercial exploitation of petroleum in 1859 to roughly the year 2000, the inflation-adjusted (to year 2000) price of oil averaged roughly $20 per barrel in most years. A barrel of oil contains the energy equivalent of roughly 23,000 hours of human labor, so $20 per barrel translated to a minuscule energy cost as compared to the cost of the energy sources (principally, human and animal muscle-power) that had built the pre-20th-century agrarian world. The industrial cities of today were founded on ultra-cheap fossil energy. Parking lots were paved, bridges spanned, suburbs and highways constructed—all with a principal fuel source that cost only an insignificant fraction of the cost of human labor.

During the past decade, the yearly average price of oil has jumped to over $100 per barrel. Global annually averaged crude oil prices doubled from $25 in 2002 to $55 in 2005, and then doubled again, from $55 in 2005 to $111 in 2011. The energy of oil is still cheap when compared to the cost of human labor, but it has increased roughly 500% in comparison to its price during the 20th century, the heyday of industrial expansion. We are still wealthy compared to our ancestors; we still enjoy the benefits of cheap energy. Yet now, paving parking lots, spanning bridges, and constructing suburbs and highways costs significantly more than it did previously.

Alternative, renewable energy sources have the potential to replace oil in some applications. Still, the inevitable energy transition away from fossil fuels will take enormous investments, and it will also take time— three or four decades in the best case. It is therefore highly unlikely that society will make sufficient investment, in sufficient time, to avert a steep decline in available energy and a steep increase in energy costs during the coming decades.

The economy can adjust to higher energy prices over time, but that adjustment process may be painful. Since replacing oil with other energy sources will be difficult, and since oil is so pivotal to world trade, the decline of oil will probably ensure the commencement of a historic period of economic contraction—in some respects, a mirror image of the 20th century's unprecedented boom.[9] And that's a fair interpretation of what we are beginning to see take place around us. Economic weakness plagues the world's industrialized nations. Efforts to extract "extreme" fossil fuels have taken on an air of desperation. The oil-rich Middle East is in turmoil, with major world powers seeking either to buy influence with rulers or to gain control of resources by destabilizing regimes. Paradoxically, while labor productivity rose during the era of ultra-cheap energy as workers used powered machines to accomplish more tasks, rising energy costs now translate to higher unemployment and downward pressure on wages.

Much of our current economic dilemma has to do with debt—and this in turn also relates to the underlying energy problem. As American economist Robert Gordon has documented, cheap oil and electrification drove rapid economic growth during the mid-20th century.[10] By the 1970s, the expansion of oil- and electricity-based infrastructure was reaching a point of diminishing returns in terms of its ability to keep the economy expanding: most families already had a car or two, as well as a houseful of electric appliances and gadgets. As globalization took hold, American factory workers found themselves competing with workers in poorer nations, and real hourly wages stopped growing. With demand stagnating, new ways had to be found to keep the engine of economic growth humming. Since the 1970s, growth in consumption has been maintained to an ever-greater degree simply by borrowing, with rising consumer debt as a significant driver of commerce. During this period in the United States, debt (all debt, not just government debt) rose at three times the rate of GDP growth. As debt ballooned, the financial industry increased in size relative to manufacturing, agriculture, and the other components of the economy. The financial industry then began blowing bubbles as a way of increasing profits. The most recent of these was the US housing bubble, whose collapse in 2007–2008 left us where we are now. The end of the era of cheap oil and the inflation and collapse of history's biggest debt bubble are historically intertwined.

With energy literacy, an understanding of energy history, and a peak-oil-informed perspective, current economic and geopolitical events become much more readily understandable—if no less causes for concern.

DEPLETIONIST ECONOMICS

Standard economic theory says that peak oil should present no problem. If any resource that is in high demand becomes scarce, its price will rise until someone finds a substitute. After all, petroleum itself was initially a substitute for whale oil. If crude oil can no longer be produced at the rate at which the world wants to consume it, we will adapt. We'll drive electric cars. We'll burn biodiesel made from algae.

Yes, we will find substitutes for oil—at least in some instances. But there is no guarantee they will be superior, or even affordable. All the substitutes currently available are problematic in one way or another.

There's a school of thought that says (in effect) that the more money we spend on energy substitutes, the better. If we spend lots of money on new energy sources (borrowing the money, if necessary), doing so will increase GDP—which is essentially a measure of how much money is spent in the economy. And a higher GDP is assumed to translate to a higher standard of living.

This line of thought may be misleading. While we tend to think of money as the prime mover of the economy, in fact it is energy that gets things done. More and cheaper energy translates to a more complex society with a growing economy; less energy, and more expensive energy, translates to a stagnant or shrinking economy that sheds complexity.[11]

Again, the primary implication of peak oil is an end to economic growth as we have known it during the past few decades. In order to adapt to peak oil, we will need not just different energy sources, but transformations in the ways we use energy. We will be less mobile and will need to adapt our trade dependencies and redesign our cities and our lives accordingly. We will need to rethink our food systems to make them more locally based and less dependent on petrochemical inputs. We will need a new economic paradigm in which growth is no longer the goal, one in which conservation of natural resources is a much higher priority than is currently the case.

In short, peak oil turns out to be a very real problem, and a very big one indeed.

★ ★ ★

All of the trends discussed above—the steep rise in oil production costs in recent years, the leveling off of world crude oil production rates, the economic pain that is resulting, and the implications for future economic growth—constitute the "game" that fracking is attempting to change. How much, and for how long, does it change that game? Let's take a look at what has been accomplished and what's been promised with this new technology.

SNAKE BITES

① ## THE INDUSTRY SHILLS SAY:
Thanks to new technologies we have a *100-year supply of natural gas here in the United States.*

THE REALITY IS:
That 100-year figure is arrived at by extrapolating results from the very best wells to entire regions and ignoring future demand trends. It's grade-A snake oil.

② ## THE CONVENTIONAL WISDOM SAYS:
Fracking is a "game changer" for domestic oil and gas production. "We can drill our way to energy independence in the United States."

THE REALITY IS:
Shale gas and tight oil, like all fossil fuels, are finite resources. The *rate of supply* from both will rapidly decline in the near future. If we don't develop long-term renewable energy alternatives now, we will be caught short.

Hydraulic fracturing and horizontal drilling have increased production of natural gas and oil by tapping vast shale deposits. But the industry has made **extraordinary claims** *about the extent and longevity of the shale boom—claims that the evidence does not support.*

Chapter Two

TECHNOLOGY TO
THE RESCUE

Fracking will end America's reliance on imported oil. The United States can look forward to a hundred years of cheap natural gas. The US will soon become energy independent and will surpass Saudi Arabia to become the world's foremost petroleum producer.

These are extraordinary claims, but they are not entirely without basis. To quote a gas exploration company representative's repeated assertion at a presentation I attended in 2009, "The proof is in the production."[1] Just a few years ago, US natural gas production was declining and apparently set to go off a cliff. Instead, today's gas-in-storage is at or near record highs, the price of natural gas has recently retreated to historic lows, and there is serious talk of exporting liquefied natural gas to other nations by ocean tankers.

Similarly, US oil production—which had been generally declining since 1970—is now on the rise, primarily because of the application of hydraulic fracturing and horizontal drilling in tight reservoirs. In 2012, US oil production soared by 766,000 barrels per day, the biggest one-year boost ever; domestic production is at its highest level in 15 years.

Nevertheless, claims that have recently been made for the potential of fracking technology to produce spectacular amounts of shale gas and tight oil for decades to come have drawn skeptical responses from some geologists. The boom has been going on for a few years now—long enough to generate data and to permit reasonable observers to gain some perspective.

In this chapter, we will review the long history of the technology behind the shale gas and tight oil booms in the United States, and the short history of the booms themselves. Then, in Chapter 3, we'll drill into data to see whether the facts really support the industry's claims.

A BRIEF HISTORY OF FRACKING

The essential purpose of hydrofracturing is to create and maintain fractures in oil- or gas-bearing rock; these fractures enable oil or gas to migrate toward a well bore so it can be extracted from the ground.

The idea of fracturing rock to free up hydrocarbons goes back almost to the beginning of the oil industry. In 1866, US Patent No. 59,936 was issued to Civil War veteran Col. Edward Roberts, who developed an invention he titled simply, "Exploding Torpedo." Roberts would lower an iron cylinder filled with 15 to 20 pounds of gunpowder into a drilled borehole until it reached oil-bearing strata. The torpedo was then exploded by means of a cap on top of the shell connected by wire to a detonator at the surface. Roberts also envisioned filling the well bore with water to provide "fluid tamping" to concentrate the concussion and more efficiently fracture the rock.

The invention worked. The Roberts Petroleum Torpedo Company went on to "shoot" thousands of Pennsylvania oil wells with explosives, and production from the wells increased as much as 1,200% within the first week after the procedure. Roberts's contracts with well owners gave him a royalty of 15% of subsequent oil production; understandably, many drillers wanted the benefit of "shooting" but not the cost, so they built their own torpedoes, exploding them at night with no observers around—a practice that gave rise to the term "moonlighting."

In the 1940s, Floyd Farris of Stanolind Oil and Gas studied the use of water as a fracturing agent, carrying out the first hydraulic fracturing experiment in 1947 at the Hugoton gas field in southwestern Kansas. His experiments led to the first commercial application of hydrofracturing in 1949, when a team of petroleum production experts applied it to an oil well near Duncan, Oklahoma. Later the same day, Halliburton and Stanolind successfully fractured another well near Holliday, Texas. Starting in the 1970s, the use of hydrofracturing became widespread within the petroleum industry, often in efforts aimed at "enhanced oil recovery" (EOR) in conventional oil and gas fields. However, oil- and gas-bearing shale rocks remained mostly out of bounds for drillers.

In the 1980s and 1990s, George P. Mitchell of Mitchell Energy & Development, now part of Devon Energy, discovered that shale has naturally occurring cracks. Some shales are more fractured than others; if hydrofracturing could be applied where cracks are already present, large amounts of gas might easily be released.

In 1991, Mitchell pioneered the use of horizontal drilling for natural gas, guiding wells down a kilometer or so, then bending the well bore to extend horizontally another kilometer. This accomplished two things: it provided more contact between the well bore and oil- or gas-bearing strata, and it allowed producers to drill horizontally beneath neighborhoods, schools, and airports—which would prove to be a great advantage in cases like the Barnett shale, where significant gas deposits lie beneath the City of Fort Worth.

A few years later, Mitchell developed "slick-water" fracturing, which involves adding friction-reducing gels to water to increase the fluid flow in fractured wells. Mitchell then combined horizontal drilling and slick-water hydraulic fracturing, and focused his efforts on producing gas from the Barnett formation in Texas.

Over the following years, the industry worked to develop more complex mixtures of fracturing fluids with ingredients including fine sand and a laundry list of chemicals, many of them toxic. Some of these materials (such as sand) act as "proppants," which are injected after the rock is initially fracked in order to prop open the newly created rock fractures. Other ingredients perform a range of functions, from optimizing fluid flow, to scouring the inside of the well casing. The exact formulas for fracking fluids are typically proprietary and carefully guarded. Changes to the Clean Water Act in the Energy Policy Act of 2005 exempted natural gas drillers from having to disclose the chemicals used in hydraulic fracturing, thus averting costly regulatory oversight. This came at the urging of then-Vice-President Dick Cheney, and the relevant passage in the Act has come to be known as the "Halliburton loophole," since Cheney had a long-standing business association with Halliburton, and that company stood to benefit substantially from the exemption.

The last key technological component of modern fracking consisted of multi-well pad or cluster drilling—the drilling of up to 16 wells from one industrial platform. This enables operators to concentrate machines and material in one place so as to reduce costs and accelerate well approvals. Cluster drilling from one pad was not introduced until 2007.

In many respects, the industry's newfound ability to access shale gas and tight oil pivots on these technological developments. But there is more to the story. Mitchell Energy's focus on unconventional gas was partly motivated by the federal government's removal of natural gas price

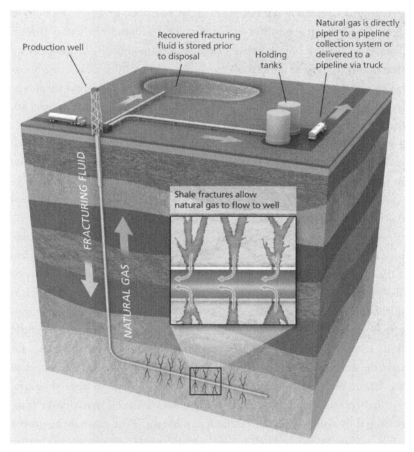

Figure 14. Schematic Diagram of a Horizontal Shale Gas Well. Multiple horizontal shale gas wells are often drilled from a common platform, with each well stimulated with multiple hydraulic fracture treatments.

Source: Image Copyright (c) The Analysis Group, 2011. Used with permission.

controls and by new federal tax credits designed to promote the development of unconventional natural gas resources. In the late 1980s and early '90s, limits to US conventional natural gas supplies were becoming apparent—limits that would lead to steeply rising gas prices in the early 2000s. The US federal government and some states began offering tax credits or severance tax abatements to companies developing tight gas, coalbed methane, or shale gas. Soaring oil prices were similarly instrumental to the development of the Bakken tight oil play. In retrospect, it's clear that it was the bringing together of several technological innovations in the context of high oil and gas prices and changes in government regulations that made large-scale commercial exploitation of shale gas and tight oil reservoirs possible.

HOW TO FRACK A SHALE GAS (OR TIGHT OIL) WELL

Suppose you want to get in on the fracking game. Here's a short instruction manual to get you going.

Start with a geological survey. You need to know where the gas or oil is, and you will probably wish to operate within one of the "plays" already identified by the industry (such as the Marcellus, Eagle Ford, or Bakken). But you need more than the general information that you can glean from the US Department of Energy and US Geological Survey websites—you need to know the location of "sweet spots" within these plays where production will be highest. You will be able to obtain that knowledge only by purchasing proprietary drilling data from other companies, and by drilling your own test wells. Recent technological innovations in 3-D seismic imaging will help immensely in enabling you to visualize exactly where the most prospective rock layers are.

Sooner or later you will need drilling leases—rights, purchased from landowners, to exploit subsurface mineral resources. Start with a search of land ownership records at county offices. Actual lease negotiations and signings may take place on doorsteps or kitchen tables in rural homes (as in the film *Promised Land*). You may want to load your boiler-plate agreement with language that allows you to build roads, buildings, gates, drilling pads, and pipelines anywhere on the owner's land; to interfere with farming, hunting, timber rights, conservation programs, and other land uses; to take millions of gallons of water from wells on the land; to leave the landowner liable for any damages caused to neighbors by your drilling practices; and to store wastewater and chemicals on the land. You'll offer the property owner an up-front bonus payment per acre (from five hundred to several thousand dollars, depending on a variety of factors), plus royalties that promise a percentage of the value of oil or gas that's produced. The lease will give you a three- to five-year deadline to drill. If a well is drilled, the lease stays in effect for as long as the well produces.

Once you know where you want to drill and you have a leasing agreement in hand, you're ready to get to work. Plan the drilling site—and, if you're drilling for gas, the pipeline route by which to move your product to market. Send some workers with earth-moving equipment to clear an area for the drilling operations: you'll need an earthen berm enclosing a football-field-sized site. The drilling rig itself—a 120-foot-high steel structure of platforms surrounding a huge rotary drill—can be rented and assembled from about 60 tractor-trailer-loads of equipment.

Drilling will probably take two or three weeks, with steel pipe being lowered into the hole as the drill bit chews its way straight down a mile or two, then turns laterally to drill outward another few thousand feet. You'll cement special steel pipe, called *casing,* into place in the uppermost parts of the well. This will protect groundwater and stabilize the well for the next stages of the process.

You're now ready to slide a device known as a "perforating gun" down to the deepest portion of the well; this sets off small explosive charges that punch holes in the horizontal steel production casing. Once that's accomplished, it's necessary to flush the system with diluted acid to unclog the holes.

Now comes the hydrofracturing stage. Bring in huge pumps on semi-trucks, along with four to six hundred tanker loads of water and fracking fluids. With the pumps, first drive a few million gallons of water mixed with "slickening" agents down into the horizontal leg of the casing, forcing the water through the holes to make hairline cracks in the shale. Then add microscopic grains of sand to the water to prop the cracks open.

After the well is fracked, you will "pump back" water and fracking fluid for several days to open up the well bore so that oil or gas can flow out. You may recapture the fracking fluid for reuse in the next job, or you might decide to put it in an evaporation pond, or send it off to a municipal treatment facility (which is probably poorly equipped to deal with it).

If you've been drilling for gas, you will now cap the well until you've constructed a pipeline to connect it with larger transmission pipes. If it's an oil well, you may be able to start production right away and move the product by truck and rail tanker.

Now it's time to drill the next well on your pad; its horizontal leg will point in a different direction from the first well. Once several wells have been drilled and you've finished with the pad, simply break down the rented drilling rig so its owner can truck it away to the next site. Most of your work is done.

As soon as you've opened the tap and started production from your new oil or gas well, you will also rehabilitate, as best you can, most of the land around the drilling site, leaving (if it's a gas well) a fenced area the size of a large living room with several pipes protruding about three feet from the ground, along with a couple of small tanks.

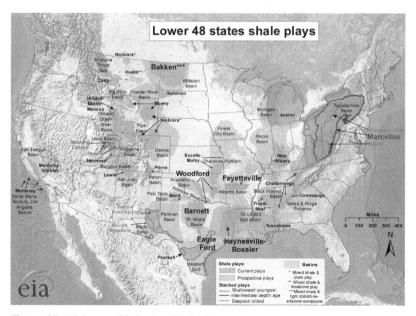

Figure 15. US Lower 48 States Shale Plays.

Source: Energy Information Administration, September 2011.

Along the way, you will have had to move a lot of equipment, water, and chemicals. Altogether, each well will have generated 1,800 to 2,600 18-wheel-truck trips.

Hiring personnel, renting the drilling rig, paying for the lease, hiring trucks—all of this is expensive. By the time you turn on the tap, you probably will have invested $10 to $20 million in your well pad—which, if you've been drilling for gas, may produce only $6 to $15 million worth of product over its lifetime at today's prices. If it's an oil well, you are more likely to show a profit, though there's no guarantee.

So why does anyone bother? That's another story—one we'll explore in Chapter 5.

THE SHALE GAS BOOM, PLAY BY PLAY

Meanwhile, let's continue with our history of the recent and ongoing fracking boom. That history is dotted with the names of the "plays," or geologic formations, where fracking is common. It takes only a few moments to grasp the essential information about each one.

As already noted, the boom got its start with the **Barnett** formation in the 14 counties in and around Dallas and Fort Worth, Texas. In the early 20th century, geologists had identified thick, black, organic-rich shale in an outcrop close to the Barnett Stream, which gave the play its name. But shale is hard and impermeable, so efforts to produce gas in commercial quantities from the formation came to little until the late 1990s. Mitchell Energy began development of the Barnett in 1999; subsequent operators have included Chesapeake, EOG Resources, Gulftex Operating, Devon Energy (which bought out Mitchell), XTO, Range Energy Resources, ConocoPhillips, Quicksilver, and Denbury. The Barnett is now dotted with nearly 15,000 gas wells, which are mostly concentrated in a "core" area of production in and close to Fort Worth, where the shale is thicker and yields more gas per well. Current production is 5.85 billion cubic feet per day, but production rates have hit a plateau since late 2011, despite an ongoing increase in the number of operating wells. (All well and production numbers cited in this chapter are accurate to June 2012.)

Development of the **Fayetteville** formation (near Fayetteville, Arkansas) began in 2002 by Southwestern Energy. A surface outcrop of organic-rich shale had been identified before 1930, but once again natural gas extraction efforts were delayed until the arrival of high prices and new technology. After confirming commercial levels of gas in the formation in 2002, Southwestern embarked on a huge and successful concealed leasing operation, securing 455,000 acres in the prime development area prior to drilling its first publicly announced "discovery" well. By late 2004, up to 25 other companies had joined the land-rush, including SEECO, Chesapeake, Petrohawk, XTO, David H. Arrington, and One-Tec (Chesapeake eventually sold its interests in the Fayetteville shale to BHP Billiton Petroleum). The area of production is spread over 25,000 square miles in parts of Cleburne, Conway, Faulkner, Jackson, Johnson, Pope, Van Buren, and White Counties and includes 3,873 wells yielding a total of 2.8 billion cubic feet per day. The recent production trend has been flat despite continued drilling, which suggests that this play is in its late-middle-age phase.

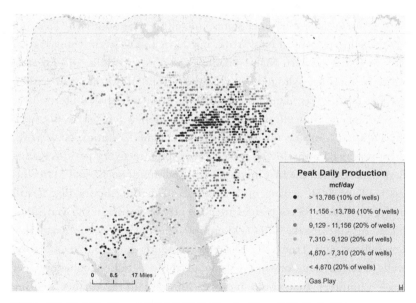

Figure 16. Distribution and Peak Daily Production of Wells in the Haynesville Shale Gas Play.
Source: Data from DI Desktop/HPDI, compiled by J. David Hughes, September 2012.

The *Haynesville* play, which straddles the Louisiana-Texas border, is named after the town of Haynesville in Claiborne Parish, Louisiana. Chesapeake was first on the scene here in early 2008, followed by

Anadarko, Petrohawk, XTO, Exco, EnCana, J-W, EOG, and SWEPI. The leasing rush and subsequent production boom have minted more than a few new millionaires in the Shreveport, Louisiana region. The Haynesville play extends under the core Texas counties of Harrison, Panola, Shelby, and San Augustine, as well as De Soto, Red River, and Caddo Parishes in Louisiana. It has an estimated 250 trillion cubic feet of recoverable gas. In 2010 another rich natural gas reservoir, the Bossier shale, was discovered overlying the Haynesville by six to eight hundred feet. There are currently over 2,800 wells operating in the formations, producing just under 7 billion cubic feet of gas per day, roughly a quarter of all US shale gas being brought to market. While production from the Haynesville formation is the highest of any US shale gas play, it is now declining: this is a fully mature play, though it is only about five years old.

The *Marcellus* play underlies a large area of the Appalachian region of the northeastern United States, including the Southern Tier and Finger Lakes regions of New York, northern and western Pennsylvania, eastern Ohio, western Maryland, most of West Virginia, and extreme western Virginia. Altogether, it covers several times more area than the Barnett. It was named for a distinctive organic shale outcrop near the village of Marcellus, New York. Though a few gas wells were drilled a half century ago in Tioga and Broome Counties, New York, these produced only slowly, with a long capital recovery period. Range Resources drilled the first modern hydrofractured, horizontal Marcellus gas well in 2004, setting off a leasing and drilling boom that is still under way. Over 85 companies are currently operating in the Marcellus, including Chesapeake, XTO, Marathon, Phillips, Chevron, Anadarko, Longfellow, and True Oil. The play currently hosts 3,850 operating wells with current total production of about 5 billion cubic feet per day—a number that could increase substantially as further drilling occurs, and especially if New York State's fracking moratorium is lifted. The Marcellus is still a youthful play.

The *Utica* shale is located a few thousand feet below the Marcellus shale and may have the potential to become a commercial natural gas resource on its own. It is thicker than the Marcellus and more geographically extensive, reaching much of Kentucky, Maryland, New York, Ohio, Pennsylvania, Tennessee, West Virginia, and Virginia, as well as parts of

Lake Ontario, Lake Erie, and Ontario, Canada. So far, the only areas of the Utica that have been subject to leasing and drilling are in eastern Ohio and Ontario, Canada, where the formation is closer to the surface and the Marcellus is not present. Wherever the Marcellus formation is present, it is the preferred production target because it is closer to the surface and thus less expensive to drill. Current production from the Utica, from a mere 13 wells, is negligible.

The *Eagle Ford* play starts at the Texas-Mexico border in Webb and Maverick Counties and extends 400 miles toward East Texas; it takes its name from the town of Eagle Ford, where the shale outcrops at the surface. The play is 50 miles wide at a depth between 4,000 and 12,000 feet, with an average thickness of 250 feet, and contains both oil and gas. Petrohawk drilled the first Eagle Ford gas well with a horizontal leg and hydraulic fracturing in 2008 in La Salle County, Texas, but dozens of operators are currently active in the play, including Chesapeake, Devon Energy, Lewis, EOG, XTO, Statoil, and Talisman. Current gas production from the Eagle Ford consists of 2.14 billion cubic feet per day from 3,129 wells.

The *Woodford* play in Oklahoma saw minor gas production as early as 1939; by late 2004 there were 24 gas wells operating, and by early 2008 that number had grown to more than 750. The largest gas producer from the Woodford is Newfield Exploration; other operators include Devon Energy, Chesapeake, Cimarex, Antero, St. Mary, XTO, Pablo, Petroquest, Continental, and Range Resources. Production from the Woodford shale has peaked and is now in decline, with 1,827 wells currently producing a total of a little over a billion cubic feet of gas per day.

There are several plays currently producing at a lower rate, such as the *Antrim* shale in Michigan (9,409 wells yielding 290 million cubic feet per day with a declining trend), as well as formations that may have some future potential but are currently yielding negligible production—including the *Caney* shale in Oklahoma, the *Conesauga* and *Floyd* shales in Alabama, and the *Gothic* shale in Colorado.

Outside the United States, shale gas resources in China exceed even those of the United States; potential exists also in South America, Europe, extreme northern and southern Africa, and Australia. However, none of these regions is currently a significant producer. (In Australia, hydrofracturing is used to produce coalbed methane; a major controversy over environmental impacts is erupting there in response.)

Altogether, in just the last decade the US shale gas industry has drilled over 60,000 wells, with a total current rate of production of about 28 billion cubic feet per day. US shale gas production appears to have peaked or leveled off in late 2011 for reasons we will explore in Chapter 3. The drilling boom has produced roughly 20 trillion cubic feet of gas—over a hundred billion dollars' worth of product—and has led to the creation of hundreds of thousands of jobs (however temporary) for drillers, truckers, and miscellaneous service personnel. Thousands of households have benefitted financially from lease and royalty payments. Utilities are now burning less coal and more natural gas due to the drilling boom. And makers of US energy policy envision more of the same—cheap, abundant natural gas for as far as the eye can see.

TIGHT OIL BY THE NUMBERS

Sometimes the shale rocks that yield natural gas to hydraulic fracturing also (or instead) contain oil. But calling this resource "shale oil" creates confusion because the term is so similar to "oil shale"—a phrase customarily applied to an entirely different resource in the western part of the United States. (We'll discuss the potential of oil shale in Chapter 6.) In order to avoid this confusion, geologists usually call crude oil that's present in shale (or similar rock) "tight oil." Conventional, commercially accessible oil is typically found in porous rock that is topped by an impermeable "cap rock" that keeps the oil from migrating to the surface and oxidizing. But that's typically not where the oil formed; it migrated there from "source rock," usually shale, which formed slowly from sediments on ancient seabeds. Tight oil is petroleum that remained in this shale source rock, kept there by unusually tight pore spaces and a lack of pathways between pores. As with shale gas, tight oil presents a challenge for drillers—one that has been partially overcome by the use of new technology.

Most production of tight oil (sometimes called "light tight oil," or LTE) in the United States is occurring in two formations—the Bakken play in Montana and North Dakota (also in Saskatchewan, Canada), and the Eagle Ford play in south Texas.

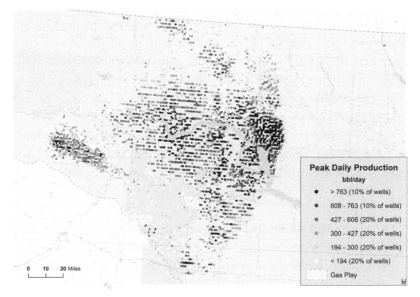

Figure 17. Distribution and Peak Daily Production of Wells in the Bakken Tight Oil Play.
Source: Data from DI Desktop/HPDI, compiled by J. David Hughes, September 2012.

Geologist J. W. Nordquist first described the **Bakken** formation in the Williston Basin in 1953, following an initial oil discovery at the Clarence Iverson farm in North Dakota in 1951. In 2000, Lyco Energy drilled the first horizontal pilot well into the Bakken; five years later EOG Resources demonstrated that horizontal drilling combined with hydraulic fracturing could recover significant oil from the play. Commercial production using modern fracking technology began in 2008, and by the end of 2010, oil production rates had reached 458,000 barrels per day, outstripping the industry's capacity to ship oil out of the region.

Repeatedly throughout the past several decades, geologists have sought to estimate the total oil endowment of the Bakken formation. A 2008 report issued by the North Dakota Department of Mineral Resources suggested that the North Dakota portion of the Bakken contains 167 billion barrels of oil—which, if it represented oil *reserves,* would put North

Dakota ahead of Iraq in terms of oil endowment. However, the percentage of this oil that can practically be extracted is highly debatable, with estimates as low as 1%. Companies operating in the Bakken include Anglo Canadian, Concho, Abraxas, EOG Resources, Continental, Whiting, Marathon, QEP, Brigham, Hess, Samson, and Statoil. As of mid-2012 there were about 4,600 wells producing a total of 570,000 barrels per day. While overall production from the Bakken has grown rapidly in recent years, the Montana portion of the play is already in decline. We will explore the overall potential for this play in the next chapter.

A *New York Times Magazine* article ("North Dakota Went Boom" by Chip Brown, January 31, 2013) chronicled the economic and social impact on life in North Dakota as a result of the leasing and drilling frenzy in the Bakken:

> It has minted millionaires, paid off mortgages, created businesses; it has raised rents, stressed roads, vexed planners and overwhelmed schools; it has polluted streams, spoiled fields and boosted crime. It has confounded kids running lemonade stands: 50 cents a cup but your customer has only hundreds in his payday wallet. Oil has financed multimillion-dollar recreation centers and new hospital wings. It has fitted highways with passing lanes and rumble strips. It has forced McDonald's to offer bonuses and brought job seekers from all over the country—truck drivers, frack hands, pipe fitters, teachers, manicurists, strippers.

Meanwhile, back in south Texas, the **Eagle Ford** play has seen substantial production of tight oil as well as shale gas (wells in the southeastern, deeper side of the play yield mainly natural gas while wells on the northwestern, shallower side yield mostly oil). Oil reserves are estimated at 3 billion barrels. Mid-2012 production was 424,000 barrels per day from 3,129 producing wells, with an increasing production trend.

While the Bakken and Eagle Ford together account for over 80% of current US tight oil production, there are other plays that offer varying degrees of promise. The **Granite Wash** formation, straddling the northern Texas-Oklahoma border, produces roughly 41,000 barrels per day from 3,090 active wells with a rising trend. The **Cline** shale, located east of Midland, Texas, in the Permian Basin, produces about 30,000 barrels per day from 1,541 wells; here again, production is increasing. Tight oil is also being produced from the **Barnett** shale in Texas, where 14,871 wells yield only 27,000 barrels per day. In this play the production trend is flat.

The **Niobrara** formation in Colorado and Wyoming presents problems with complex geology and access to water, especially given the severe drought that has gripped much of the United States, and Colorado's recent catastrophic wildfires. Early comparisons with the Bakken have not borne out, and disappointing well results have led Chesapeake to sell off its Colorado leases. Noble, Anadarko, EOG, Quicksilver, and roughly a dozen other mostly small companies are competing in the play, with about 40 active drilling rigs. However, 10,811 operating wells currently yield a mere 51,000 barrels per day of production, and the production trend is flat.

The **Austin Chalk** play (which reaches across Texas and into Louisiana) and the **Spraberry** play (near Midland, Texas) each produce over 17,000 barrels per day.

The **Monterey** shale in Kern, Orange, Ventura, Monterey, and Santa Barbara Counties in southern California boasts tight oil resources of up to 15 billion barrels—four times the size of Bakken reserves. But resources are not the same as reserves, and so far production amounts to only 8,580 barrels per day from 675 operating wells, with a flat production trend. This could change if drilling picks up, in view of the Monterey's very high resource endowment.

Elsewhere in the world, geology appropriate for the production of tight oil using fracking technology exists in R'Mah formation in Syria, Sargelu formation in the northern Persian Gulf region, Athel formation in Oman, Bazhenov formation and Achimov formation of west Siberia in Russia, Coober Pedy in Australia, and Chicontepec formation in Mexico. Little is yet being done to exploit these resources.

THE CLAIMS RUSH

It may be helpful to pause at this point and recall again where we were at the start of the fracking boom. US oil production had generally been in decline for nearly four decades, oil and gas prices were high and rising, and mainstream media outlets were beginning occasionally to mention the possibility that world petroleum output was near its inevitable peak. In this context, rising gas production from north-central Texas, Arkansas, Louisiana, and Pennsylvania, and soaring oil yields in North Dakota and south Texas seemed like answers to a prayer. Here was an opportunity for the industry to beat back its critics—and make a lot of money in the process.

The situation recalls events in the 1970s. Oil price shocks during that decade, along with declining US oil production, provoked discussion about ultimate limits to petroleum supplies. America experienced a natural gas crisis as well: wellhead prices jumped more than 400% between 1971 and 1978, while production declined more than 11%. The oil dilemma was resolved by new discoveries in Alaska and the North Sea: petroleum prices declined in the 1980s and stayed low for over a decade. The US natural gas market was eventually rebalanced by demand destruction, with reduced consumption leading to stable and affordable prices that would last, again, until the early 2000s. Throughout the late 1980s and the 1990s, cheap oil and stable, affordable gas prices enabled Americans to forget about the need for energy conservation and the development of renewable energy sources, and to concentrate on their favorite pastimes—driving and consuming. Might the fracking boom offer similar relief to the oil and gas price spikes of the 2000s? The industry obviously thought so, and it was determined to make the most of the opportunity.

But there was a more immediate, practical motive for oil and gas companies to ballyhoo fracking's significance: their need for investment capital. Small operators willing to assume substantial risk by developing marginal resource plays using expensive technology have led the fracking boom from its inception. These companies need investors to believe that fracking is the Next Big Thing. As in every resource boom since the dawn of time, hyperbole has become a tool of survival.

The hurricane of hype began in the shale gas fields of Texas, stirred by the charismatic Aubrey McClendon, then-CEO of Chesapeake Energy. McClendon hammered home the same message on every possible occasion—at investment conferences, in government hearings, and in prominent media interviews. For example, in testimony before the US House Select Committee on Energy Independence and Global Warming on July 30, 2008, McClendon had this to say:

> America is at the beginning of a great natural gas boom. This boom can largely solve our present energy crisis. The domestic gas industry

through new technology has found enough natural gas right here in America to heat homes, generate electricity, make chemicals, plastics and fertilizers, and most importantly, potentially fuel millions of cars and trucks for decades to come.

Another highly visible shale gas booster was Daniel Yergin, chairman of Cambridge Energy Research Associates, an oil and gas industry consultancy. In an April 2, 2011, article in the *Wall Street Journal* titled "Stepping on the Gas," Yergin wrote: "Estimates of the entire natural-gas resource base, taking shale gas into account, are now as high as 2,500 trillion cubic feet, with a further 500 trillion cubic feet in Canada. That amounts to a more than 100-year supply of natural gas."

A century of natural gas! It was a nice round figure, and big enough to banish any fears of looming scarcity. The number came to be repeated so frequently that even President Barack Obama parroted it unquestioningly, as in this public statement on January 25, 2012: "We have a supply of natural gas that can last America nearly 100 years, and my administration will take every possible action to safely develop this energy."

But was one hundred years really enough?

Oil billionaire T. Boone Pickens, whose hedge fund had adopted significant positions in the natural gas sector starting in 1997, began running a series of television and print advertisements in 2008 to promote his "Pickens Plan" to "break the stranglehold of imported oil" using domestic natural gas for transportation. In an interview on CNBC in April 2011, he estimated America's natural gas endowment: "If I announced that we have more oil equivalent than the Saudis do, I would be telling you the truth. . . . I say you're going to recover 4,000 trillion [cubic feet]. Which is 700 billion barrels." It turns out that 4,000 trillion cubic feet (tcf) is roughly the equivalent of 160 years of US natural gas production at current rates.[2]

A hundred years? 160 years? Why not more? So far, Aubrey McClendon appears to have topped all rivals with his claim, in an article on Chesapeake Energy's website, that America has *two hundred years* of natural gas.[3] In his most widely heard prediction about the importance of shale gas, in a CBS News *60 Minutes* interview that aired on November 14, 2010, McClendon told Leslie Stahl: "In the last few years we have discovered the equivalent of two Saudi Arabias of oil in the form of natural gas in the United States. Not one, but two." As if betting in a poker game,

McClendon seemed to be saying, "I'll see your Saudi Arabia and raise you one! And I'll double down on that 'hundred years,' too!"

As we will see in more detail in the next chapter, even Daniel Yergin's seemingly conservative hundred-year estimate is unsupportable and overstates supplies by several hundred percent. How could McClendon, Yergin, and Pickens possibly have come up with these super-optimistic shale gas supply forecasts? Simply by taking the highest imaginable resource estimate for each play, then taking the best imaginable recovery rate (based on extrapolating data from the very best-producing wells in the small "sweet spots" in each play), then adding up the numbers. Always the assumption was that the gas could be produced profitably at current prices. Only the most knowledgeable experts would know that the resulting figures were entirely unrealistic.

Fast-moving developments in the shale gas sector came as a surprise to official agencies like the US Department of Energy's Energy Information Administration (EIA), the United States Geological Survey (USGS), and the International Energy Agency (IEA). None of these agencies had foreseen that high gas prices would lead small producers to apply fracking technology to known shale plays, and with such spectacular results. The EIA quickly sought to catch up to the industry's achievements—in both production and public relations—by issuing new forecasts for future shale gas production. Borrowing uncritically from the gas producers' own estimates, the EIA assigned a reserves figure of 410 trillion cubic feet to the Marcellus play alone. Soon the USGS weighed in, suggesting the real figure should be closer to 84 tcf; the EIA quickly backtracked and deferred to the USGS, cutting its own estimate for the Marcellus by 80%.[4] The episode simply served to illustrate that ostensibly authoritative reserves and future production forecast numbers were in fact highly speculative, with enormous error bars.

Meanwhile, public perceptions about the prospects for tight oil followed a similar trajectory. Early resource claims for the Bakken play were all over the map. A research paper by USGS geochemist Leigh Price in 1999 had estimated the total amount of oil contained in the Bakken shale at somewhere between 271 and 503 billion barrels.[5] Later estimates by Meissner and Banks (2000) and by Flannery and Kraus (2006) ranged all the way from 32 to 300 billion barrels.[6]

If the amount of oil in place was a matter for dispute, the question of how much of this was recoverable constituted an even more decisive unknown variable. Here the estimates ranged from as little as 1% to as much as 50%. The USGS currently estimates the Bakken to have 3.65 billion barrels of technically recoverable oil in place (the more crucial *economically* recoverable amount is likely substantially lower).[7] That's still a big number, but it represents only six weeks of current world oil consumption.

Again, the industry, in its public statements, focused only on the largest numbers for both resources-in-place and recovery potential. The Bakken and Eagle Ford were heralded as the biggest developments in the oil world since the invention of the drill bit. Everyone involved would get rich, the boom would last decades, and it would lead America's energy sector into a new Golden Age of plenty.

The industry's PR efforts received an enormous boost from Leonardo Maugeri, senior manager for the Italian oil company Eni and senior fellow at Harvard University, who published a seemingly authoritative paper in June 2012 titled, "Oil: The Next Revolution."[8] In it, Maugeri claimed that "The shale/tight oil boom in the United States is not a temporary bubble, but the most important revolution in the oil sector in decades." Published under the imprint of Harvard's Kennedy School, Belfer Center for Science and International Affairs, the Maugeri report painted a euphoric picture of world oil abundance: "Oil is not in short supply. From a purely physical point of view, there are huge volumes of conventional and unconventional oils still to be developed, with no 'peak-oil' in sight."

At the center of this portrait of abundance was US tight oil. While Maugeri managed to identify a few other promising places such as Iraq—where production, he figured, could go from the current rate of 3.35 million barrels per day to over 5 mb/d by 2020 (a highly optimistic notion, given the political realities there)—he saved his biggest hopes for the Bakken, Eagle Ford, and other North American tight oil plays. One phrase from the report leapt out: ". . . *the total production capacity of the US could even exceed that of Saudi Arabia.*" According to Maugeri, the United States could get an additional 4.17 million barrels per day from tight oil plays by the end of the decade. To put that number in perspective, total US production of crude oil in 2011 was 5.68 million barrels per day. Adding 4.17 mb/d to that number would yield a total almost equal to America's peak level of production achieved in 1970 and also close to Saudi Arabia's current production of about 10 mb/d. Energy reporters, taking their cue from Maugeri, began adopting the shorthand term, "Saudi America."

Maugeri's report received uncritical notice in major media outlets, including the *New York Times*, the *Wall Street Journal*, NPR, and most broadcast and cable news television networks, and his assertions became common wisdom. This happened despite the presence of several pivotal and fairly obvious errors in the report, including Maugeri's consistent confusion of "depletion rate" with "decline rate," a serious underestimation of decline rates from existing oil fields, and a simple but decisive math mistake in compounding declines.[9] It turned out that the report had not been peer-reviewed or even competently fact-checked. "Oil: The Next Revolution" was thoroughly debunked by experts, but none of the criticisms surfaced in publications that had turned the report into headline news. It wasn't hard to see why: Maugeri's twisted tune was music to the ears of the oil industry.

Official agencies began revising their oil reserves numbers and production forecasts. The EIA, in its "Annual Energy Outlook 2013," noted that US oil import dependency had fallen from 60% of total oil consumed in 2005 to 45% in 2011; assuming further growth in tight oil output, the agency projected oil imports to fall to only 37% of consumption in 2035. The United States would not achieve oil independence, but it would make substantial progress in that direction.

The IEA likewise adopted a more optimistic attitude about future petroleum supplies. The organization's chief economist, Fatih Birol, even called the surge in US oil and gas production "the biggest change in the energy world since World War II."[10]

As with shale gas, Daniel Yergin played a key role in pumping up expectations about the potential of tight oil. "[T]echnology has opened doors people didn't know were there or didn't think could be opened," he told the *Wall Street Journal*. "We expect to see tight-oil production grow dramatically over the rest of this decade."[11]

★ ★ ★

Altogether, these were amazing developments. Prior to the fracking boom, the United States had been assumed to be a fully mature oil and gas province. Since the start of the hydrocarbon era, more oil wells had been drilled in the continental US than in all other countries combined. The nation's peaks in oil and gas production were apparently four decades in the rearview mirror. Yet, led by technology and enabled by the treatment of mineral rights under US property law, a host of small oil and gas companies had unleashed a genie of new production potential.

Nevertheless, there was another way of framing the situation. Soaring fuel prices, resulting from the depletion of giant conventional fields, had led drilling companies to go after some of the last, least inviting oil and gas plays in North America. These operators had invented superior barrel-scrapers, but they were still in essence scraping the bottom of the barrel by producing oil and gas from source rocks.

Every geologist understands the principle of the *resource pyramid*: the entire pyramid represents the total mineral resource in place. The top portion of the pyramid consists of the concentrated, easy-to-produce portion of that resource base, while lower levels correspond to more abundant but lower-quality resources that have higher production costs and whose extraction implies higher environmental risks. This mental model holds

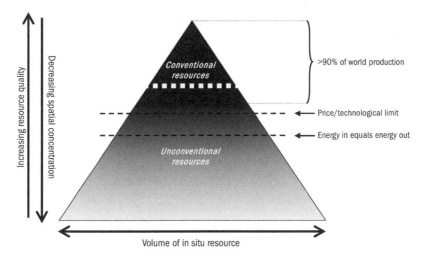

Figure 18. Oil and Gas Resource Volume Versus Resource Quality. This graphic illustrates the relationship of in situ resource volumes to the distribution of conventional and unconventional accumulations, and the generally declining net energy and increasing difficulty of extraction as volumes increase lower in the pyramid.

Source: J. David Hughes, "Drill, Baby, Drill," Figure 37.

true for copper and iron mines, oil and gas fields, and even commercial fisheries. Shale gas and tight oil plays were far from the top of America's gas and oil resource pyramids. In addition, each shale gas or tight oil play could be thought of as its own smaller resource pyramid: the best resources within each play would inevitably be targeted for production first, and, as time went on and as producers made their way down the stair steps of the pyramid, well productivity would decline and per-well decline rates would rise. Operating costs would soar. Production potentials that were forecast on the basis of extrapolating the best results from the first wells drilled into "sweet spots" in each play would inevitably prove highly misleading.

But not many analysts wanted to adopt this more realistic view. There was no money in it.

Leonardo Maugeri's statement that "the shale/tight oil boom in the United States is not a temporary bubble" carries a whiff of resemblance to Nixon's "I am not a crook" or Clinton's "I did not have sexual relations with that woman": the gentleman doth protest too much, methinks. But where lies the truth? Are shale gas and tight oil booms the "new normal" for American energy? Or do they more closely resemble a short-term Ponzi scheme?

Let's take a closer look.

SNAKE BITES

(1) THE US ENERGY INFORMATION ADMINISTRATION (EIA) SAYS:
Enough new shale gas wells will be drilled every year until 2030 to ensure steady production growth.

THE REALITY IS:
Production from shale gas wells typically declines 80 to 95 percent in the first 36 months of operation. Just to maintain the current rate of supply will take massively increased rates of drilling.

(2) THE EIA SAYS:
Proved and unproved technically recoverable shale gas reserves will provide a 24-year supply of natural gas at current US consumption rates.

THE REALITY IS:
Given steep shale gas well decline rates and low recovery efficiency, the United States *may actually have fewer than 10 years* of shale gas supply at the current rate of consumption.

*Eventually, horizontal drilling is suspended because operators reach a point where they are **just burning cash.***

~ Robert Smith, operations geologist, International Western Oil

Chapter Three

A TREADMILL TO HELL

The tiny ghost town of Desdemona is situated in Eastland County, Texas, about halfway between Fort Worth and Abilene. It was founded in the mid-19th century as a fort to protect settlers from Indians, its early economy revolving mostly around peanut farming. In 1918, Tom Dees of Hog Creek Oil Company discovered an oil field nearby, and within weeks 16,000 speculators and rig workers crowded Desdemona's dusty streets. Fortunes were quickly made—less often on actual oil production than on the trading of stock shares, which appreciated dramatically in value during the first couple of years of the boom. (Some shares that originally sold for one hundred dollars soon fetched over ten thousand.) Fortunes were just as suddenly lost in gambling or robberies. By 1920, rampant lawlessness had drawn the attention of the Texas Rangers, who at the time operated as a paramilitary organization employing tactics like targeted killing and enhanced interrogation. The Rangers effectively ran Desdemona—but they didn't stay long. Between 1919 and 1921, oil production rates dropped by two-thirds. The value of oil stocks collapsed. By 1936, Desdemona's city government had dissolved itself; the town's lone school closed its doors in 1969, and as of 2013 only two businesses remain.

Booms go bust: it is a story as old as civilization. Historically, most booms have been associated with resource extraction—gold, silver, oil, gas, or coal. Often, financial speculation based on an extravagant (and sometimes deliberate) overestimation of resource potential drives the peak of the boom higher than would otherwise be the case, thus making the bust all the more devastating. Though the pattern is consistent, on each occasion the participants assure themselves and one another that "this time it's different."

The current fracking frenzy in the oil and gas fields of Texas, North Dakota, Oklahoma, Louisiana, Arkansas, Colorado, and Pennsylvania shows all the signs of being a boom in the classic sense. How do we know it's *not* different this time, that it *won't* end in a colossal bust? And if it *is* yet another instance of the same old story, how soon will the bust come?

These are questions best answered by data—by realistic resource estimates, per-well production and decline rates, and reliable calculations of the number of possible drilling sites. Compiling these kinds of data is hard work and often requires access to expensive proprietary information. And the rewards are few: investors want good news.

In the previous chapter we surveyed the claims made by the industry regarding reserves and future production of shale gas and tight oil. This chapter tells the story of the data—how they have evolved, and what they tell us now.

THE BOOM THAT FIZZLED

The first indication that the emerging shale gas bonanza might not have a happy ending came in 2007 when Arthur E. Berman, a petroleum geologist and consultant to the oil and gas industry in Sugar Land, Texas, started crunching numbers from the Barnett shale gas play. The results were anything but encouraging. Berman used his regular column ("What's New in Exploration") in *World Oil* magazine to report on his analysis of decline rates and profitability for hydrofractured, horizontally drilled Barnett wells and concluded: "This analysis shows that, while many wells are profitable and some operators are significantly more successful than others, most Barnett shale wells will lose money. . . . The overall resource size for the play is great, but economic reserves are relatively small."

Berman continued accumulating data from shale plays and publishing his analyses; the following paragraph, from his March 6, 2009, *World Oil* column titled "Shale Plays, Risk Analysis and Other Perils of Conventional Thinking: Haynesville Shale Sizzle Turns to Fizzle," is typical of his coverage:

An early analysis of 20 horizontally drilled wells in the Haynesville Shale play in Louisiana and parts of adjacent East Texas suggests a disappointing outcome because of extremely high decline rates. Average monthly decline rates are 24%, with 75% of wells declining 20–35% per month. The impressive initial production rates (IP) for these wells do not, therefore, necessarily translate into high reserves (actual daily production rates from the maximum 30-day period were, in fact, about 20% lower than reported IPs).

Representatives of the shale gas industry hotly denied Berman's assertions, accusing him of "inconsistent data gathering" and of having "poorly supported opinions." After all, total production in the shale gas plays was rising, companies were flush with investment capital, and jobs were being created. How could this be anything less than a game-changing economic miracle?

In October 2009, Berman wrote yet another column questioning shale gas prospects; this time *World Oil* refused to run it. Berman recounted the events this way when I e-mailed him:

> Perry Fischer, the editor of *World Oil*, called to tell me that my column in press would be pulled because of objections from Petrohawk Energy and Seneca Resources. I later had a conversation with John Royall, President and CEO of Gulf Publishing that owns *World Oil,* who objected to my comments to the press that he had been pressured by industry not to publish my articles on shale. My relationship was based on freedom to choose content. Since the magazine rejected content, I chose to end the relationship.

The incident was reported in the *Houston Chronicle* (Nov. 3), which noted:

> John Royall . . . said he didn't receive any pressure from gas companies. *World Oil* serves a global audience, and gas shale is largely a domestic issue. Berman had written on the topic for a year, and Royall decided that was enough. "Art had an interesting take on shale gas," he said. "It was interesting, provocative stuff, but it was time to move on."[1]

Berman continued gathering data, doing the numbers, and writing his conclusions in articles published at TheOilDrum.com and on his own his own blog, PetroleumTruthReport.blogspot.com. He also gave public presentations, including one at the 2009 ASPO-USA conference in Denver, where I first heard him speak.

By 2011, Berman had been joined by other critics of the shale gas boom. Bill Powers, editor of *Powers Energy Investor* and previously the editor of the *Canadian Energy Viewpoint* and *US Energy Investor,* began telling his readers about high decline rates and other problems repeatedly and in detail, relying mostly on Berman's analysis. "The importance of shale gas has been grossly overstated," Powers told TheEnergyReport.com. "The US has nowhere *close* to a 100-year supply. This myth has been perpetuated by self-interested industry, media and politicians."[2] Soon Powers began working on a book, *Cold, Hungry and in the Dark: Exploding the Natural Gas Supply Myth,* published earlier this year. Art Berman contributed the book's foreword.

Berman's work also served as initial inspiration for a major new analytic survey, the most comprehensive to date, authored by David Hughes and published by the Post Carbon Institute (at which I am senior fellow) in February 2013.[3] Hughes, a geoscientist who studied the energy resources of Canada for nearly four decades, including 32 years with the Geological Survey of Canada as a scientist and research manager, examined proprietary data on 63,000 US shale gas and tight oil wells, calculating production decline rates in each active play. The data were licensed from DI Desktop. (DI stands for DrillingInfo, a petroleum industry data company headquartered in Austin, Texas.) Hughes's report fills over 160 pages, including many tables and graphs, and also addresses prospects for expanded production of tar sands and other unconventional fuels. For anyone wanting to understand current and future production from fracking and horizontal drilling, "Drill, Baby, Drill" is the Holy Grail of information and analysis. Here's the report's abstract:

> It is now assumed that recent advances in fossil fuel production— particularly for shale gas and shale oil—herald a new age of energy abundance, even "energy independence," for the United States. Nevertheless, the most thorough public analysis to date of the production history and the economic, environmental, and geological constraints of these resources in North America shows that they will inevitably fall short of such expectations, for two main reasons: First, shale gas and shale oil wells have proven to deplete quickly, the best fields have already been tapped, and no major new field discoveries

are expected; thus with average per-well productivity declining and ever-more wells (and fields) required simply to maintain production, an "exploration treadmill" limits the long-term potential of shale resources. Second, although tar sands, deepwater oil, oil shales, coal-bed methane, and other non-conventional fossil fuel resources exist in vast deposits, their exploitation continues to require such enormous expenditures of resources and logistical effort that rapid scaling up of production to market-transforming levels is all but impossible; the big "tanks" of these resources are inherently constrained by small "taps."[4]

From the work of Berman, Hughes, and other analysts, a more realistic picture of the actual potential of shale gas and tight oil plays is emerging. Briefly: The wells in *core areas* (usually just a few counties) in each of these plays do tend to be productive and profitable, yielding oil or gas in significant amounts for many years. These are not comparable to the conventional oil and gas finds of the mid-20th century, but they do nevertheless provide an important new source of supply for the industry and for the nation. However, these compact core areas tend to be drilled out fairly quickly. Meanwhile, outside these regions, per-well production rates tend to fall quickly and dramatically, and wells are uneconomic. Overall, taking into account decline rates, potential drilling locations, and the variability of regions within resource plays, *the industry's claims for how much oil and gas can be extracted, at what rate, and how profitably, are wildly overblown.*

Throughout much of the rest of this chapter, as we explore the energy reality of fracking in more detail, we will be relying primarily on data and analysis in "Drill, Baby, Drill."

SHALE GAS: THE EVIDENCE IS IN

When discussing US shale gas production, it is always necessary to begin by acknowledging the industry's accomplishments—as we have already done on more than one occasion in this book. Natural gas production in the United States is now higher than at any point in history, and shale gas currently makes up 40% of America's total natural gas production. Considering how quickly the new technology has been deployed, this is an impressive achievement.

Nevertheless, it turns out that high productivity shale gas plays are few and far between: just six plays account for 88% of total production. And, as noted at the end of Chapter 2, each play is in effect its own "resource pyramid," characterized by a few small "sweet spots" surrounded by larger areas capable of only marginal productivity. Drillers invariably concentrate their efforts on the zones of highest productivity first. So, as time goes on and as drillers must stray ever further from sweet spots, the initial productivity of each *new* well drilled in the play tends to be lower than that of previous wells. The number of available drilling sites is always limited, and, once the play is saturated with wells, per-well decline rates will determine the play's longevity.

Hughes notes that individual shale gas well decline rates range from 80–95% after 36 months, in the top five US plays. The industry's claim that America has 100 years of gas is based on the assumption that individual wells will continue to produce for 40 years, but given such steep decline rates, the data do not support this assumption.

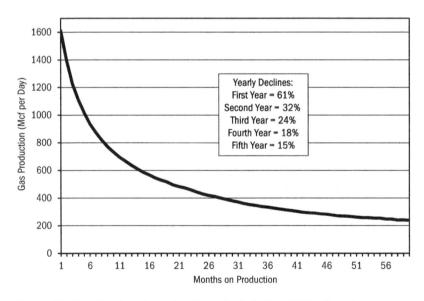

Figure 19. Type Decline Curve for Barnett Shale Gas Wells. Based on data from the most recent five years of this play's production.

Source: J. David Hughes, "Drill, Baby, Drill," Figure 48. Data from DI Desktop/HPDI current through May 2012.

One result of high decline rates is that a large proportion of overall field output must be replaced by additional drilling in order to keep the total production rate growing or even flat. Hughes calculates that, for the nation as a whole, between 30 and 50% of shale gas production must be replaced *every year* with more drilling—amounting to roughly 7,200 new wells a year. Remember: that's simply to maintain the current production rate. This is the "treadmill to hell" referred to in the title of this chapter. Oil analyst Rune Likvern uses a different metaphor; he calls it the "Red Queen" syndrome, after a character in Lewis Carroll's *Through the Looking-Glass*. In that colorful story, the fictional Red Queen jogs along at top speed but never gets anywhere; as she tells Alice, "It takes all the running you can do, to keep in the same place." Similarly, with such steep decline rates, it takes all the drilling that the industry can do just to keep production steady.[5]

There were 341,678 operating gas wells in the United States in 2000, prior to the fracking revolution, representing more than a century of drilling efforts. In 2011, that number had swollen to 514,637.[6] Here again is evidence that descent to lower levels of the "resource pyramid" ensures diminishing returns from increasing effort: since 1990 the average productivity per well has declined by 38%.

The EIA reports these trends but still believes shale gas production rates can continue to grow. What would it take to make that happen? Only a drilling pace that's utterly unprecedented can possibly suffice. In the 2005–2008 period, the industry roughly tripled the number of natural gas wells being drilled annually, as compared to 1990s' rates. To produce the estimated US reserves of shale gas, the EIA calculates that 410,722 shale gas wells will have to be drilled.[7] It takes a moment to mentally process the implications of drilling on that scale.

Obviously, this would represent an enormous, unprecedented investment on the part of the gas industry. Already, dry shale gas plays require $42 billion per year in capital investment in drilling in order to offset declines. Given current low natural gas prices (as of this writing, natural gas is selling for about $4 per million Btus), this investment is not recouped by sales: in 2012, US shale gas generated just $33 billion in revenues. As we'll see in more detail in Chapter 5, gas drilling companies are staving off bankruptcy through a variety of strategies, including asset sales and increased production of liquid fuels. How realistic is it to assume that these companies will double down on their dry gas drilling investments during the next couple of decades, absent much higher gas prices?

And what do these trends suggest about the reliability of shale gas reserves numbers? Clearly, shale gas *resources* do exist in enormous quantity. But *reserves* are always a fraction of the total resource base. Some reserves are termed *technical reserves:* these are resources that theoretically could be extracted given current technology. A smaller but more important category consists of *economic reserves:* these are resources that can profitably be extracted with current technology and at current prices. If the industry is, on the whole, losing money on shale gas production, this suggests that US economic reserves of shale gas are in fact fairly modest. At higher prices, more resources would fall into this category. If gas prices were $15 per million Btus (as they already are in some parts of the world) instead of $4, then economic reserves would grow accordingly. But the American people are being led to believe that most of the shale gas resource base can be produced at a price low enough so as to enable natural gas to be used for the majority of power generation, and even as a substitute for gasoline in tens of millions of cars and trucks. This is pure folly.

Finally, what are we to make of the familiar claim that the United States is sitting on a hundred years' worth of natural gas? It is clearly not based on realistic public data. The EIA lists proved and unproved technically recoverable shale gas reserves at almost 600 trillion cubic feet (tcf). This is 24 years of natural gas supplies at current US consumption rates. But even this 24-year supply estimate is questionable. David Hughes notes: "This is an extremely aggressive forecast, considering that most of this production is from unproved resources, and would entail a drilling boom that would make the environmental concerns with hydraulic fracturing experienced to date pale by comparison."[8]

Rafael Sandrea of IPC Petroleum Consultants, in a report titled "Evaluating Production Potential of Mature US Oil, Gas Shale Plays," notes that unusually high field decline rates associated with shale gas plays imply low recovery efficiencies. "The average recovery efficiency is about 7%," he writes, "in contrast to recovery efficiencies of 75–80% for conventional gas fields. This suggests that the estimate of recoverable gas for all US shale plays should be near 240 tcf."[9] Which is less than 10 years of current United States natural gas consumption.

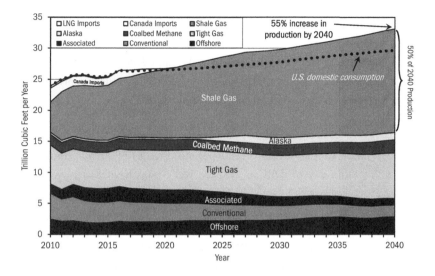

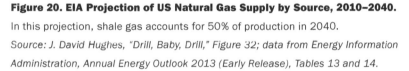

Figure 20. EIA Projection of US Natural Gas Supply by Source, 2010–2040.

In this projection, shale gas accounts for 50% of production in 2040.

Source: J. David Hughes, "Drill, Baby, Drill," Figure 32; data from Energy Information Administration, Annual Energy Outlook 2013 (Early Release), Tables 13 and 14.

BAKKEN BOOM, BAKKEN BUST

The situation we've just surveyed with regard to shale gas is largely mir-rored in the tight oil plays of North Dakota and south Texas. Again, per-well production decline rates are steep—between 81 and 90% in the first 24 months. Production from individual wells tapers off so quickly that 40% of overall output (from older wells with lower decline rates along with output from newer ones) must be replaced annually by new drilling just to keep the total supply curve flat. According to Hughes, "Together the Bakken and Eagle Ford plays may yield a little over 5 billion barrels—less than 10 months of US consumption."[10]

The Bakken play had produced 0.5 billion barrels through May 2012, with an estimated ultimate recovery of about 3 billion barrels by 2025. On one hand, this represents a remarkable accomplishment: who in 2000 or even 2005 expected North Dakota to become a major oil-producing region? Yet the achievement requires extraordinary effort. Drillers can't let up; if they do, high per-well decline rates will ensure falling overall production.

An article by Jaci Conrad Pearson in the *Black Hills Pioneer* (September 19, 2012) titled "It Takes Oil Money to Make Oil Money" captures the expense of an enterprise involving hundreds of companies and thousands of wells:

> "It takes $3 per second, $180 per minute, $10,800 per hour and $259,000 a day to drill an onshore well," said Kent Ellis, owner of Aurora Energy Solutions, LLC, an oil and gas brokerage firm with offices in Bismarck, ND and Oklahoma City, Oklahoma, during his address to a crowd of more than 100 gathered for his presentation as part of the Black Hills Pioneer's Oil, Gas and Mineral Rights Workshop. ". . . It takes 2,200 gallons-plus of diesel fuel a day, just to run the rig." And moving the rig is another story and another significant cost. "To move a rig from Spearfish to Belle Fourche costs around $250,000," Ellis said.[11]

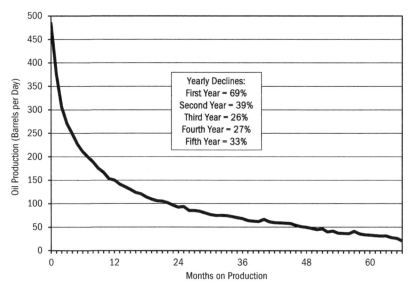

Figure 21. Type Decline Curve for Bakken Tight Oil Wells. Based on data from the most recent 66 months of this play's oil production.

Source: J. David Hughes, "Drill, Baby, Drill," Figure 63; data from DI Desktop/HPDI current through May 2012.

This is not your grandfather's oil business. Tight oil deposits are typically thinner than those in conventional wells, with layers of oil-bearing rock sandwiched between other rock layers. Horizontal drilling enables the operator to go after oil deposits from the side, yielding much higher recovery than a vertical well could achieve. But it also implies a dramatic production decline curve. In effect, operators must chase the deposit sideways, and the cost of drilling horizontally in pursuit of the ever-retreating reserve quickly escalates. "Eventually," according to Robert Smith, operations geologist with International Western Oil, "horizontal drilling is suspended because operators reach a point where they are just burning cash."[12]

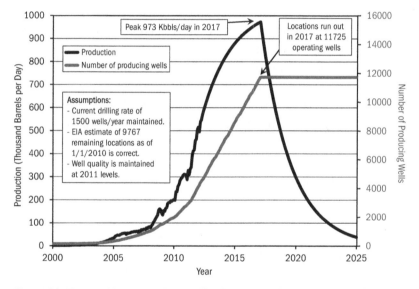

Figure 22. Future Oil Production Profile for the Bakken Play, Assuming Current Rate of New Well Additions. Based on data from the most recent 66 months of this play's oil production.

Source: J. David Hughes, "Drill, Baby, Drill," Figure 66; historical data from DI Desktop/HPDI current through May 2012.

The Eagle Ford is younger in its production cycle than the Bakken. Operators there are still in the process of identifying sweet spots; while they find and drill these optimum locations, average initial production rates are still rising. Still, Eagle Ford decline rates are even higher than those observed in the Bakken. The first-year decline in production in new Eagle Ford wells is 60% and the overall decline at the end of the second year is 89% below the average initial production levels of wells drilled in 2012. These decline rates mean the average Eagle Ford well will enter the category of "stripper" well (yielding fewer than 15 barrels per day) within about three years.

For every play there are only so many places to drill. For the Eagle Ford, the EIA estimates a total of 11,406 effective locations. With a 40% overall field decline rate, and assuming current rates of drilling with all new wells performing as in 2011, Hughes anticipates a peak of production in the Eagle Ford in 2016 at 0.891 million barrels per day.[13] Total oil recovery is estimated at about 2.23 billion barrels by 2025, amounting to a five-month contribution to US oil consumption.[14]

More than 80% of current tight oil production in the United States comes from the Bakken and Eagle Ford, with the other 20% issuing from 19 other formations. Estimates suggest the biggest prize of all could be the Monterey shale in California, with 41% of America's total purported tight oil resources. But Hughes is not optimistic about the Monterey play's prospects: "Recent drilling results have been disappointing and the longer-term performance of the Monterey is mostly at 'stripper well' levels . . . with an average of 12.7 barrels per day from [each of] 675 wells. This bears no comparison to the Bakken or Eagle Ford."[15] The problem is geological: California's seismic history has left the Monterey shale heavily faulted, folded, and fractured, presenting drillers with far more expensive complications than ones they face in North Dakota and Texas.[16]

The United States' total tight oil "technically recoverable unproved resources" are estimated at between 23 and 34.6 billion barrels (assuming that 13.7 billion barrels can be produced from the Monterey play). "Although significant," writes Hughes, "this is hardly cause for celebrating US 'energy independence,' as it represents somewhere between three and four years of consumption, even if it all could be recovered—which would take decades."

WILL THE REST OF THE WORLD GET FRACKED?

Some fracking boosters claim that the United States is merely the thin end of a wedge, and that the same technology that opened up the Barnett and

Bakken will soon liberate oil and natural gas from similar reservoirs in China, Europe, and elsewhere. How likely is this?

The US fracking boom is several years old now, and so far little shale gas or tight oil production is occurring in other parts of the world. This could simply be a problem of timing: perhaps the rest of the world will eventually catch up with North America. On the other hand, there could be fundamental barriers to the widespread application of fracking technology outside the United States. Let's explore the factors at work and see whether they support an expectation of worldwide shale gas and tight oil abundance.

Some countries have banned, tightly regulated, or put off fracking for environmental reasons. Outright bans have been enacted in France, Luxembourg, and Bulgaria. In Germany, Poland, and the United Kingdom, tight regulations constrain drillers. Throughout most of Europe there is strong public opposition to fracking on environmental grounds. Whether these are temporary or persisting impediments to industry development will depend on forthcoming revelations about the environmental safety of fracking, and on industry efforts to address the problems. As we'll see in Chapter 4, the impacts to air quality, water quality, and climate from shale gas and tight oil production are hardly trivial.

In the United States, public opposition to fracking has been attenuated by the system of private ownership of mineral rights. Households that stand to gain thousands, perhaps even a few million dollars, from leasing drilling rights and from subsequent production royalties are often willing not just to overlook environmental problems, but to actively oppose other members of their communities who seek to enact drilling moratoria or bans. In most other nations, the government owns all mineral rights. Local environmental problems that ensue from fracking are therefore likely to provoke much more local opposition outside the United States.

While large shale gas and tight oil reserves numbers are often touted for other nations, those numbers are highly speculative. According to a 2011 US Energy Information Administration estimate, Poland has Europe's largest recoverable reserves of shale gas—187 trillion cubic feet, a third more than those of the Marcellus shale. However, this is a fairly meaningless statistic: the Polish Geological Institute estimates the nation's reserves at 27 trillion cubic feet, only about one-seventh the EIA figure. Until many wells have been drilled and are in production, both numbers are mere guesses. Currently, nearly 1,200 drilling rigs are busy perforating America's shale beds; Poland so far deploys only half a dozen rigs.

Hence another problem with the worldwide deployment of fracking: the lack of technology. The oil and gas industry got its start in the United States, and America has always enjoyed a technological edge when it comes to drilling. Most of the world's oil services companies, which pioneered nearly all of the important innovations in drilling during the past century and a half, are headquartered in Texas. The United States has half the world's drilling rigs, and American colleges and universities still turn out the bulk of the world's petroleum geologists and engineers. Other countries—China comes quickly to mind—could make the enormous investments required to develop the needed technology, build the rigs, and train the experts. But it would still take time.

Water can also be a limiting factor. Saudi Arabia has plenty of gas-bearing shales, but little of the water that would be necessary to hydro-fracture them. As climate change brings more extreme periodic drought conditions to nations like Australia and China, high water demands may make hydrofracturing problematic-to-impossible in those countries as well.[17]

Geology is a problem too. As we've seen, not all US shale plays are created equal, and even in the best of them only localized "core areas" are actually profitable to drill. The same principle holds for the rest of the world. China's shale gas resources are purported to be the world's largest, beating even those of the United States. But the Chinese formations are more complex than those in the United States, many having a high clay content, which makes them more pliable and less apt to fracture. Also, many Chinese shale plays are deeper, requiring higher per-well investment in drilling. Compounding these problems, China lacks means for compiling, assessing, and sharing geological data comparable to those developed during the past century in the United States.[18]

Financial factors also constrain non-US development of shale gas and tight oil. In the United States, the fracking boom was driven by small companies willing to take on substantial financial risk. The industry was also buoyed by investor capital funneled by Wall Street, which tirelessly hyped the nation's prospects for a century's worth of cheap gas and oil. In many other countries, state-owned companies do the drilling, and investment decisions are made by risk-averse bureaucrats rather than risk-seeking, hype-driven capitalists.

Altogether, the evidence suggests that other nations *are* working to develop the means to extract shale gas and tight oil resources, and that they *will* eventually have some success. But the process will take years, and there is no nation in which the oil and gas industry is likely to fully repeat the performance of the independent companies operating in Texas, North Dakota, Pennsylvania, Louisiana, and Arkansas.

WHY DO OFFICIAL AGENCIES SO OFTEN GET IT WRONG?

The picture we've been painting in this chapter is radically different from the one that fracking boosters portray. But it differs also from the forecasts of official agencies—principally, the International Energy Agency and the US Energy Information Administration—and from those of oil industry sources such as BP's annual "Statistical Review of World Energy." Reading David Hughes's "Drill, Baby, Drill" report, one encounters statements like these:

> The IEA's suggestion that these costs will not escalate further over the next 23 years, as assumed in its $10 trillion upstream oil forecast, seems wishful thinking indeed. (p. 26)

> The growth in shale/tight oil production in this projection is very aggressive, requiring the consumption of 26 billion barrels, or 78%, of the EIA's estimated unproved technically recoverable shale/tight oil resource by 2040. The likelihood of this happening is remote. (p. 34)

Why should we believe Hughes but not the EIA? Does the agency ever get its numbers wrong? Yes, as a matter of fact, it frequently does. Hughes provides a graphic of 12 EIA forecasts for world oil production going back to 2000, noting, "Compared to actual 2011 production, these projections invariably overestimated world oil production levels. The 2002 projection, for example, overestimated 2011 production by 13%, or 11 [million barrels per day]—and that was only nine years out."[19]

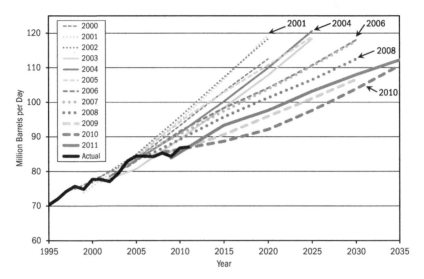

Figure 23. EIA World Oil Production Estimates Compared to Actual Production, 2000–2011. Most cases overestimated actual 2011 production. *Source: J. David Hughes, "Drill, Baby, Drill," Figure 25. Data from Energy Information Administration.*

In March 2012, the EIA published an "Annual Energy Outlook Retrospective Review: Evaluation of 2011 and Prior Reference Case Projections," in which it found that during the past dozen years it had underestimated oil prices and overestimated oil production most of the time. (More specifically, the agency found that it had overestimated crude oil production 62% of the time and overestimated natural gas production 70.8% of the time.)[20]

There is evidence to suggest that this pattern of poor forecasting is ongoing. Roger Blanchard, author of *The Future of Global Oil Production: Facts, Figures, Trends and Projections by Region,* notes that recent EIA reports assume that US offshore oil production will continue to increase over time. However, the agency's own data show that total Gulf of Mexico oil production achieved its highest level in 2009 and has declined every year since then.

Table. U.S. Offshore Oil Production, Energy Information Administration Estimates Versus Actual, 2009–2012. Estimates are from *Annual Energy Outlook 2010.*

Year	EIA Estimated Production (mb/d)	Actual Production (mb/d)	Difference between EIA Estimates and Actual Production (b/d)
2009	N/A	1.62	N/A
2010	1.67	1.61	60,000
2011	1.77	1.37	400,000
2012	1.82	1.29**	530,000

** Data through October 2012.

Source: Roger D. Blanchard, "Commentary: US DOE/EIA Forecast Estimates Face Reality," ASPO–USA (website), January 14, 2013, http://aspousa.org/2013/01/ commentary-us-doeeia-forecast-estimates-face-reality/.

The IEA has a record that's no better. The 2000 IEA forecast for the price of oil a decade hence, adjusted for inflation to the 2000 dollar, was $28.25. The actual price in 2010 was $79.61, roughly *three times* the forecast price (and that was in the wake of the Great Recession). Also in 2000, the IEA forecast that total world liquid fuels production would reach 95.8 million barrels per day in 2010. The actual figure was 87.1 mb/d.

More examples could easily be cited—including ones from BP's annual "Statistical Review of World Energy." The natural question: Why have these agencies apparently adopted a bias toward overestimating production and underestimating prices? Longtime observers tend to agree that the agencies do not intend to deceive; they are merely producing demand-driven forecasts arrived at by assuming continuous GDP growth and a corresponding increased requirement for energy. Geological limits and the need for capital investment play a minor role in these calculations.

In 2009, the *Guardian* reported that a senior IEA official accused the agency of deliberately underplaying a looming world oil shortage for fear of triggering panic buying. "The senior official claims," according to the story, that "the US has played an influential role in encouraging the watchdog to underplay the rate of decline from existing oil fields while overplaying the chances of finding new reserves."[21] The International Energy Agency, set up in the 1970s to warn the world's industrialized nations about future oil shocks, evidently bows to pressure from the United States. Meanwhile, the US Department of Energy's Energy Information Administration appears to make its forecasts of future oil production conform to politically comfortable assumptions about economic growth.

During the past decade, there has been one notable exception to the agencies' tendency toward over-optimism: in the years prior to 2009, the EIA and IEA failed to foresee the substantial increase in US natural gas and oil production resulting from the application of hydrofracturing and horizontal drilling. But as soon as the new trend of growing production in Texas, North Dakota, and Pennsylvania became apparent, the agencies appear to have overcompensated. They quickly reverted to their usual pattern of overestimating future supplies and underestimating future prices.[22]

THE BOTTOM LINE ON FRACKING'S POTENTIAL TO REVOLUTIONIZE OIL AND GAS PRODUCTION

Raymond Pierrehumbert, Professor of Geophysical Sciences at the University of Chicago, recently summarized the situation with crystalline brevity: "Oil production technology is giving us ever more expensive oil with ever-diminishing returns for the ever-increasing effort that needs to be invested."[23] The numbers tell the story: in the decade between 1994 and 2004, roughly $2.4 trillion in oil industry capital expenditures buoyed the worldwide rate of oil production by 12 million barrels per day. Yet a similar $2.4 trillion in capital expenditures spent from 2005 to 2010 failed to stem the tide of declining production in the world's older, supergiant oil fields. Global oil production during those five years *declined* by two hundred thousand barrels per day.[24] The ongoing substitution of conventional, cheap oil with expensive, technology-intensive, unconventional oil sources can be compared to the human body's use of internal energy resources in the absence of sufficient food. If one doesn't eat for a few days, the body starts burning stored fat, then muscle, and finally tissues surrounding internal organs. Each next step in the process reduces overall health but is necessary to maintain life. Similarly, the global economy naturally prefers to burn regular, conventional, cheap petroleum. But as supplies dwindle, markets prioritize the use of functionally similar fuels, even though their extraction requires much higher rates of drilling, is therefore more expensive (thus impacting oil prices and the global economy), and is more environmentally risky.

Fracking gives our current energy system a brief, fragile reprieve. New extraction technology cannot return us to the bygone era of cheap energy and easy economic growth. The best it can do is to buy us a few years of relative economic stability in which to develop alternative energy sources and build low-energy transport and food systems.

But instead of embarking on that needed project, our political leaders have unquestioningly seized on exaggerated claims from oil industry hucksters promising a century of cheap natural gas and soaring oil production rates. The result is an "all of the above" energy policy with no clear direction, and a dangerous complacency about the fate of essential but highly vulnerable food and transport systems designed during past decades of hydrocarbon abundance.

Rather than a century of plenty, we face the likely recommencement of declines in US oil and gas production *before 2020*. We've purchased a few years of respite from the relentless and inevitable erosion of our nation's oil and gas production rates, but at what cost?

SNAKE BITES

THE INDUSTRY SHILLS SAY:
Hydraulic fracturing technology has a strong environmental track record.

THE REALITY IS:
- Fracking consumes millions of gallons of freshwater, pollutes groundwater and air, and—thanks to leaking methane—may contribute more to climate change than burning coal.

- An EPA report demonstrated that fracking wastewater is too radioactive to be dealt with safely by municipal treatment plants.

- One study showed that average methane concentrations in water wells near active fracking sites were 17 times higher than in wells in inactive areas.

- Fracking can lead to ozone pollution, which inflames lung tissues, causing coughing, chest pain, and asthma. Children and the elderly suffer the most.

*The spread of fracking has led to a nationwide backlash of protests led not by big environmental organizations but by **ordinary citizens** who are seeing serious impacts to water and air quality, public health, livestock, and wildlife.*

Chapter Four

FRACKING WARS, FRACKING CASUALTIES

News item, dateline February 14, 2013: Ben Lupo, 62, owner of Hardrock Excavating in Poland, Ohio, was charged with violating the Federal Clean Water Act by ordering an employee to dump thousands of gallons of brine and fracking waste discharge into a tributary of the Mahoning River. Lupo faces up to three years in prison, a $250,000 fine, and a year of supervised release if convicted. He has pleaded innocent.

Fracking opponents in Ohio seized upon the Lupo incident to call for a ban or moratorium on drilling. Fracking supporters insisted this was merely an isolated case; further, they said, the fact Lupo was caught and prosecuted simply showed that existing regulations were sufficient and effective.

It would be reassuring to know the Lupo incident did indeed represent a unique or rare occurrence, and that fracking is otherwise as safe as a walk in the park. The oil and gas industry, after all, claims to be making serious attempts to address environmental problems as they arise—finding better ways to dispose of or recycle wastewater, building better well casings, and exploring methods of capturing fugitive methane.

But fracking by its very nature implies a wide range of environmental risks, of which failure to properly treat wastewater is only one. Remember: as society extracts fuels from lower and lower levels of the resource pyramid, it must use ever more extreme measures, and more things can go wrong. Further, as we have just seen, the high per-well decline rates associated with shale gas and tight oil wells mean that drillers must frack relentlessly in order to maintain production rates; therefore environmental risks are multiplied thousands, tens of thousands, and ultimately hundreds of thousands of times over.

Across America, hundreds of grassroots groups with names like New Yorkers Against Fracking, Save Colorado from Fracking, Blackfeet Anti-Fracking Coalition, No Frackin' PA!, Don't Frack Ohio, and Ban Michigan Fracking have sprung up and formed mutual support networks. Many of the people who start or join such groups had never previously thought of themselves as environmentalists but are compelled to action by methane in drinking water, sickened livestock, bad air quality, or constant truck noise.

In response, the industry has mounted a public relations offensive. The pro-fracking website energyfromshale.org insists, for example, that "hydraulic fracturing technology has a strong environmental track record" and that "properly designed and constructed oil and natural gas wells present low environmental risk to our groundwater."

Why has there been such a massive grassroots backlash against fracking? In this chapter, we'll look at the evidence for fracking's impacts on water, air, land, and climate. Reader warning: it ain't pretty.

WATER

Everyone agrees that fracking takes water—lots of it. A single well-pad cluster might require more than 60 million gallons. Where does all this water come from? Sometimes drillers buy water from wells on leased property, sometimes they pump it from nearby streams or rivers, sometimes they purchase it from municipal water systems. In the dry states of the American southwest, future drilling could draw water from the Colorado River at a rate equivalent to that of an additional large city, yet the region already faces the prospect of serious water shortages.[1] As climate change results in more widespread and severe drought conditions, finding water for shale gas and tight oil production is likely to pose an ever more vexing conundrum. One arid county in New Mexico has already banned fracking due to its fierce water needs.[2]

That's only the start of fracking's water problems. After water has been injected deep underground in the hydrofracturing process, most of it is pumped back to the surface. At that point, the water carries with it not only a secret cocktail of chemicals added so that it can accomplish its mission, but also highly corrosive salts, carcinogenic benzene, and radioactive elements like cesium and uranium, all leached from rock strata miles underground.[3]

What's a fracker to do with all this toxic wastewater? There are several options. Drillers can inject it into deep wells—either older abandoned oil or gas wells, or holes newly drilled for the purpose. Wastewater can also be held in large evaporation pools or sent to municipal treatment facilities. Each of these options is problematic. Underground injection simply means taking precious freshwater out of aquifers or rivers, polluting it, and then burying it so that it can never be used again. Evaporation pools poison birds and are prone to leaks and spills. Municipal water treatment plants are poorly equipped to remove the pollutants in fracking wastewater, especially when many of those pollutants are company secrets. An additional problem for wastewater treatment plants is the radioactivity released in fracking: reports from the US Environmental Protection Agency (EPA) made public in 2011 showed that fracking wastewater is too radioactive to be dealt with safely by municipal treatment plants, raising the specter of entire cities drinking radioactive water so that residents can continue burning natural gas.[4]

Increasingly, fracking operations recycle most of their water, using wastewater from one well in the next well's initial hydrofracturing. This helps with the problems of sourcing water for operations and disposing of waste, but it is far from a complete solution. While the industry says it is aiming for 100% recycling, that goal is probably unattainable for purely practical reasons; currently, recycling efforts achieve about 50% efficiency. New sources of water are still needed, and toxic effluents have a way of leaking and seeping.

In October 2011, the EPA announced plans to develop standards for disposing of fracking wastewater; as of this writing, those standards have yet to be issued.

Fracking wastewater can make its way into streams and rivers, impacting both municipal water supplies and wildlife. A study published in the *Proceedings of the National Academy of Sciences* documents how chloride from fracking wastewater ends up in Pennsylvania's rivers and streams, even when the wastewater has been treated at municipal facilities.[5] The same study also found that waterways are impacted by increased amounts of total suspended solids (TSS) from shale gas drilling. High TSS levels decrease the amount of dissolved oxygen in streams, raise water temperatures, and block sunlight. The study found that 18 well pads in a watershed increases TSS concentrations by 5%. For perspective, consider that 4,000 well pads have been constructed in Pennsylvania since the beginning of the fracking boom.[6]

Shale gas drilling also runs the risk of contaminating water tables. Drillers guard against this by isolating water tables from wells with cemented-steel well casings. However, well casings sometimes fail. The industry claims that casings fail less than 1% of the time, yet independent research suggests the failure frequency may be much higher, perhaps in the range of 6–7%.[7]

Eventually (speaking now in terms of centuries and millennia) *all* well casings will leak. When a well reaches the end of its useful life, operators install cement plugs in the borehole to prevent migration of fluids between the different rock layers. This may render the well safe for decades to come, but seismic activity can dislodge even the most carefully placed plugs. According to a paper by Maurice B. Dusseault, Malcolm Gray, and Pawel A. Nawrocki, published by the Society of Petroleum Engineers in 2000, "Oil and gas wells can develop gas leaks along the casing years after production has ceased and the well has been plugged and abandoned."[8] The most frequent reason for such failures is probably cement shrinkage, leading to fractures that are propagated upward by the slow accumulation of gas under pressure behind the casing.

Once again, the high rates of drilling required in order to maintain overall field production rates in shale gas and tight oil plays serve to amplify risk: even if just 1% of well casings fail, for the more than 65,000 current wells in fracking country that translates to 650 instances of likely contamination. If failure rates are 6%, that raises the number to 3,900. Actual instances of water table pollution resulting from well casing problems are documented, despite industry efforts to deny, distract, and evade: for example, in 2007 the faulty cement seal of a fracked well in Bainbridge, Ohio, allowed gas from a shale layer to leak into an underground drinking water source; the methane built up until it caused an explosion in a homeowner's basement.[9] Other such tales would likely be more commonly heard were it not for the industry's insistence on nondisclosure agreements when landowners whose water has been contaminated settle lawsuits with drillers.

Anecdotes about flammable tap water or dying house pets can be emotionally compelling, but at the end of the day, decisions about whether to allow or ban fracking must be based on scientific studies and statistical analyses addressing the question of whether and to what degree drilling actually impacts the water we drink. Such studies have been slow to appear, partly because of industry efforts to withhold or suppress information. Nevertheless, according to one report, published in 2011 in the *Proceedings of the National Academy of Sciences,* drinking water samples from 68 wells in the Marcellus and Utica shale plays were contaminated with excess methane.[10] The study found that average methane concentrations in wells near active fracturing operations were 17 times higher than in wells in inactive areas. Methane concentrations varied according to proximity to the drilling sites. Subsequent tests confirmed the findings.[11]

While more research is needed, initial findings suggest that fracking and water safety just don't mix.

AIR

Methane, the primary constituent of natural gas, is colorless, odorless, and nontoxic—though in significant concentrations it is highly explosive. When methane is inadvertently released in gas or oil drilling, it reacts with atmospheric hydroxyl radicals (OH) to produce water to produce water vapor and carbon dioxide, which are likewise nontoxic. (We'll discuss the climate impacts of methane and carbon dioxide later in this chapter; for now we are concerned only with toxic air pollution.)

However, other chemicals often present in natural gas at the wellhead—including hydrogen sulfide, ethane, propane, butane, pentane, benzene, and other hydrocarbons—can degrade air quality significantly. In addition, emissions from trucks, compressors, pumps, and other equipment used in drilling contain a complex mixture of benzene, toluene, and xylene, as well as other volatile organic compounds. Drilling activity and truck traffic create high levels of dust, while flaring of methane also contributes to air pollution. Some chemicals associated with drilling combine with nitrogen oxides to form ground-level ozone. It is often difficult to

trace the exact causal connections between oil and gas drilling, air pollution, and human health impacts; however, people who work at or live
near fracking sites have complained of a wide variety of new illnesses with
symptoms including skin rashes, open sores, nosebleeds, stomach cramps,
loss of smell, swollen and itching eyes, despondency, and depression.[12]

Ozone pollution is normally associated with automobile exhaust,
but fracking also generates it when the volatile organic compounds in
wastewater ponds evaporate and come in contact with diesel exhaust from
trucks and generators at the well site. Ozone inflames lung tissues and can
cause coughing, chest pains, and asthma. Human health is harmed both by
prolonged exposure to low-level ozone concentrations and by exposure to
higher levels for shorter durations. Children and the elderly are the most
susceptible.[13]

Tight oil production in North Dakota releases lots of associated methane—but, given a lack of pipeline infrastructure, drillers usually just burn
the methane on-site rather than attempting to capture it. Nighttime satel-
lite photos of the state show light from natural gas flares rivaling the city
lights of Chicago and other major metropolitan areas. Flaring can result
in the emission of a host of air pollutants, depending on the chemical
composition of the gas and the temperature of the flare. Emissions from
flaring may include hydrogen sulfide, benzene, formaldehyde, polycyclic
aromatic hydrocarbons (such as naphthalene), acetaldehyde, acrolein, pro-
pylene, toluene, xylenes, ethyl benzene, and hexane. Canadian research-
ers have measured more than 60 air pollutants downwind of natural gas
flares.[14]

Once again, anecdotes are easy to come by (such as the story of Joyce
Mitchell of Hickory, Pennsylvania, who leased drilling rights on her
land to Range Resources only to endure a constant barrage of noxious
fumes),[15] but also easy to brush off as isolated incidents that don't reflect
the actual safety record of the industry. Scientific studies and statistical
analyses are crucial but have been slow to appear.

A recent article in the journal *Environmental Science and Technology*
concluded, on the basis of data from National Oceanic and Atmospheric
Administration (NOAA), that oil and gas activity in the Wattenberg field
in the Niobrara formation in Colorado "contributed about 55% of the
volatile organic compounds linked to unhealthy ground-level ozone"
in the area. NOAA maintains an air-monitoring tower outside the small

town of Erie, Colorado, located in the Niobrara play, and found that this town of 18,000 now has methane and butane spikes that exceed by four to nine times the levels of those pollutants in Dallas, Texas, a city with some of the worst air in America.[16]

A study by The Endocrine Disruption Exchange, led by environmental health analyst Dr. Theo Colborn, measured more than 44 hazardous pollutants at operating well sites in Garfield County, Colorado. The study detected pollutants up to seven-tenths of a mile from the well site. Many of the chemicals detected are known to impact the brain and nervous system; some are known hormonal system disruptors. The human endocrine system is so sensitive that even tiny doses of some of these chemicals, measured in parts per billion, can lead to large health effects.[17]

As gas drilling expands throughout the nation, production is moving closer to populated areas, with wells in some states now being drilled within a few hundred feet of schools and homes.

Expect bad air.

LAND

Suppose you own farmland, and you also own the subsurface mineral rights to your land.[18] A petroleum company offers you money for the right to drill on your property. Drillers promise to rehabilitate the acreage when they're done. You need the money. What should you do?

Your answer may depend on how badly you need the money. But it will also reflect your philosophy—whether you see the land merely as a speculative investment, or whether you have a sense of obligation to its welfare. That's because drilling for shale gas or tight oil can seriously impact land—whether through water, air, or soil pollution; damage to vegetation, livestock, and wildlife; or erosion and induced earthquakes.

Heavy metals such as lead, mercury, cadmium, chromium, barium, and arsenic have been found in soils near natural gas drilling sites. And when fracking leads to increased ground-level ozone, plants are damaged by inhibited photosynthesis and root development.[19]

Livestock and wildlife, attracted by the salty taste of fracking fluids and wastewater, can be poisoned—either dying outright or suffering loss of reproductive function, stillbirths, birth defects, and other maladies.[20]

Light and noise from fracking and related traffic can also increase animal stress. A peer-reviewed study in 2012 by professor of molecular medicine Robert Oswald and veterinarian Michelle Bamberger found significant adverse health links between fracking and livestock exposed to the air and water by-products of drilling. Animals were found to suffer neurological, reproductive, and gastrointestinal disabilities.[21]

Colorado Division of Wildlife officials have observed both indirect effects leading to population declines and direct mortality in wildlife, in areas of intensive natural gas drilling. Waterfowl are most directly impacted, as they can land in wastewater pits near drilling pads. Industrial activity too near an area where a raptor pair is engaged in courtship behavior may discourage mating, and stress can cause the adult birds to abandon eggs or even young; the loss of a breeding season reduces population over time. Deer and elk populations decline when areas of unbroken habitat are reduced by land fragmentation from road building and well pad construction. Wild fish populations are threatened by spills of industrial materials or toxic chemicals.[22]

In mountainous regions of the Marcellus shale formation, drilling leads to erosion. Loosened sediments quickly enter surface streams, contaminating coldwater fish habitats and drinking water sources.[23]

Most earthquakes triggered by hydraulic fracturing are too weak to be felt or cause significant damage—though the number of quakes in normally seismically quiet parts of fracking country in Arkansas, Texas, Ohio, and Colorado is on the rise. Indeed, according to a USGS study, in the last four years the number of quakes in the middle of the United States jumped elevenfold from the previous three decades. The largest yet measured, in central Oklahoma on November 6, 2011, was a magnitude 5.7 temblor tied to the injection of fracking wastewater. It was the biggest quake ever recorded in Oklahoma, destroying 14 homes, buckling a highway, and leaving two people injured.[24] Earthquakes are an especially serious issue in California, a state riddled with seismic faults, yet also contemplating whether to allow drillers to increasingly exploit the Monterey shale formation.

The various forms of land damage from fracking often result in decreased property values, making resale and farming difficult, and also making it harder to acquire mortgages and insurance. Properties adjoining drilling sites are often simply unsellable, as no one wants to live with the noise, the bad air, and the possibility of water pollution.[25]

All of these problems are once again multiplied by fracking's need for heroic rates of drilling, and therefore for enormous numbers of drilling sites. Consider just the state of Colorado: at the start of 2012 approximately nine thousand square miles of public land in Colorado had been leased to the oil and gas industry for drilling—roughly 10% of the state. The amount of private land under lease is probably greater, though exact figures are harder to come by. Thus, it is likely that roughly a fifth of the land area of Colorado is currently leased by the oil and gas industry.

Finally, fracking can affect land far away from drilling sites. Sand is being mined in Wisconsin, Minnesota, and Iowa for fracking in Pennsylvania, Texas, and North Dakota. The round, fine-grained sand in these regions, left over from the grinding of glacier upon rock during the last Ice Age, is ideal for use as a fracking proppant, but mining operations destroy farmland, impact wildlife, and degrade streams.[26] Moreover, when winds take up the tiny silica particles dislodged by mining, higher rates of silicosis and cancer result.[27]

CLIMATE

Considering emissions only at the point of combustion, current US natural gas power plants produce 56% less carbon dioxide, per kilowatt-hour, than existing coal plants.[28] Therefore, as the world gradually transitions toward renewable energy sources, it might seem prudent to replace coal-burning power plants with natural gas burners as a stopgap measure. That way, natural gas could serve as a *bridge fuel* to reduce carbon emissions while society makes the investments and builds the infrastructure to eventually power itself with wind and solar. This line of reasoning has been so appealing to the Environmental Defense Fund, as well as to former Sierra Club Executive Director Carl Pope and New York City Mayor Michael Bloomberg, that all have gone on record as supporting fracking for natural gas. (The Sierra Club now opposes fracking.)

However, recent research challenges the assumption that shale gas is better for the climate than coal. In 2011, Robert Howarth, professor of marine ecology at Cornell University, led a study published in *Climatic*

Change concluding that as much as 1.9% of the gas in a typical well escapes to the atmosphere during fracking, compared with 0.01% in a conventional gas well.[29] This turns out to make an enormous difference: over short time frames, methane is 20 to 100 times as powerful a greenhouse gas as carbon dioxide. If Howarth's figures are accurate, this means that life-cycle greenhouse gas (GHG) emissions from shale gas are 20% to 100% higher than those from coal on a 20-year time frame basis.

Howarth's conclusions were reported in the *New York Times*[30] and immediately triggered a firestorm of vitriolic criticism from the industry—and from a few environmental organizations. A *Forbes* article later noted that "almost every independent researcher—at the Environmental Defense Fund, the Natural Resources Defense Council, the Council on Foreign Relations, the Energy Department and numerous independent university teams, including a Carnegie Mellon study partly financed by the Sierra Club—has slammed Howarth's conclusions."[31]

The critics insisted that Howarth had greatly overestimated leaks from fracked wells. National Energy Technology Laboratory (NETL) scientist Timothy Skone provided evidence for this view in a lecture at Cornell titled "Life Cycle Greenhouse Gas Analysis of Natural Gas Extraction & Delivery in the United States"; soon afterward, Lawrence Cathles, a Cornell geology professor, similarly argued—in a commentary published in *Climatic Change*—that Howarth's fugitive methane figures were *10 times* too high.[32]

How could scientists analyzing the same phenomenon arrive at such starkly different conclusions? Three main variables are keys to understanding the discrepancy. The first is the actual level of methane emissions during the drilling and fracking of a typical shale gas well. This number must be amortized over the total lifetime production of the well in question, as most of the "fugitive" methane escapes during drilling rather than during later production. Hence the second variable: the actual lifetime cumulative production figure for a typical well in the given formation. The third significant number indicates the amount of natural gas that leaks from pipelines on its way to the end user. The *total* amount of gas leaked to the atmosphere (from all phases of production and distribution) must remain below about 3.2% of all gas produced if natural gas is to have a climate advantage over coal over the next few critical decades, during which society must avert catastrophic climate change.[33] In 2012, my colleague at Post Carbon Institute, geologist David Hughes, helped clarify issues in the dispute in a report titled "Life Cycle Greenhouse Gas Emissions from Shale Gas Compared to Coal: An Analysis of Two Conflicting Studies."[34] Hughes found that Howarth's critics were lowballing per-well methane leaks during drilling *and* overestimating likely lifetime per-well production figures. He concluded that if these numbers are corrected, "the result is not significantly different from the conclusions of Howarth et al."

Some recent measurements of actual methane emissions during drilling have come down strongly on Howarth's side. A study by the National Oceanic and Atmospheric Administration (NOAA) reported that fully 4% of the methane gas being produced in the Wattenberg field in Colorado was leaking to the atmosphere; in a subsequent study, the same NOAA team found that 9% of produced gas was leaking to the atmosphere in a large natural gas field on mostly Indian land in north central Utah.[35] On the other hand, the research arm of ExxonMobil has published a study insisting that fugitive methane leaks from fracking are less than 1% of the gas produced, though this was based on a generous estimate of lifetime production from a typical Marcellus shale gas well.[36] The US Environmental Protection Agency recently downgraded its estimates of methane leaks in America's natural gas production and distribution system, but this action was based on industry-funded studies (like the one just mentioned) rather than new direct measurements, and it did not take into account the recent NOAA data.[37]

Fugitive emissions of natural gas from pipelines (the third significant number) are still relatively poorly understood. A recent study of leaks from gas pipelines under Manhattan streets yielded numbers well above previous estimates for distribution leakage; the subsequent report estimated, on the basis of measured pipeline leaks, that the total of production losses and transmission losses for natural gas used in New York City must be above 5%.

Can drilling, production, and transmission leaks be plugged? Yes, in principle.[38] And of course, industry *should* do everything in its power to reduce fugitive methane emissions. But recent data suggest that both drillers and pipeline operators have a big job on their hands.

Meanwhile, for the time being, the evidence is strong that current full-cycle greenhouse gas emissions for natural gas—*especially from fracking*—are worse than those for coal over the first 40 years. From the standpoint of climate stabilization, fracking for gas may be a bridge to nowhere.

★ ★ ★

We've been told that the economic and climate benefits of fracking (the latter in the case of natural gas, not oil) outweigh the risks to the immediate environment and to human health. But if evidence we've surveyed in the last two chapters is credible, then the real benefits of this technology have been exaggerated, and the risks substantially downplayed.

When benefits are systematically hyped and risks are unrealistically minimized, the results are bad investments and bad government policy.

SNAKE BITES

① **THE INDUSTRY SHILLS SAY:**
Fracking creates a huge number of jobs.

THE REALITY IS:
The industry has massively oversold its jobs record.
Since 2003, oil and gas jobs account for less than
1/20th of 1 percent of the overall US labor market.
Numerous industry-funded studies count strippers
and prostitutes as "new" jobs created by the spread
of fracking.

② **THE INDUSTRY SHILLS SAY:**
Shale gas promises continued economic benefits
for decades to come.

THE REALITY IS:
A peak in US shale gas production *within this
decade* will prove those promises to be
purposefully deceptive lies.

THE WALL STREET CONNECTION:
*Investment bankers love to inflate bubbles. When
they deflate, it's not the bankers' money that's
lost—**it's more likely yours.** The fracking industry
shows all the classic signs of a bubble, including a
heavy reliance on debt with revenues from
production failing to cover operating costs.*

Chapter Five

THE ECONOMICS
OF FRACKING:
WHO BENEFITS?

In the 1980s, two Oklahoma twentysomethings, named Aubrey McClendon and Tom L. Ward, pooled $50,000 of investment capital and started a natural gas company. They decided to call it Chesapeake Energy, since McClendon (who emerged as the company's moving force) was particularly fond of the Chesapeake Bay region of Maryland and Virginia. From the start, McClendon focused the business on unconventional gas plays and cutting-edge drilling technologies.

Fast-forward to the mid-2000s. As the fracking frenzy was starting to sizzle in Texas and Oklahoma, Chesapeake was there turning up the heat. With the Barnett, Fayetteville, and Haynesville plays yielding shale gas in ever-greater amounts, Chesapeake Energy quickly became America's second-biggest natural gas producer, and a Fortune 500 company with over 13,000 employees. By 2008, Aubrey McClendon was the highest-paid of all CEOs of S&P 500 companies.[1] In 2011, *Forbes* named him "America's Most Reckless Billionaire" in a cover story detailing his lavish and highly leveraged lifestyle.[2] He owned homes in several states, a mansion on "Billionaire's Row" in Bermuda, and 16 antique boats worth nine million dollars. He also had a habit of using his property as collateral in order to borrow money with which to buy still more.

Critics began drawing comparisons between Chesapeake and Enron, the energy giant whose infamous bankruptcy in 2001 made it synonymous with galactic-scale accounting fraud. On the surface, they were very different entities: while Enron was a labyrinthine organization with a black-box trading operation and a flotilla of off-balance-sheet shell subcompanies, Chesapeake had real assets in proven and unproven gas reserves and a real product to sell. Still, Chesapeake's business practices were opaque even to some of its biggest investors, and its cash flow from operations was insufficient to cover exploration and development costs and acquisitions in any of the last 10 calendar years.[3]

How was Chesapeake making money? Early in the shale boom, the company bought drilling leases at a pace unrivaled; later, it sold bundles of these leases to other operators, many of them European or Chinese, and usually at a profit. Chesapeake was also adept at entering into partnerships and joint ventures. And it used Byzantine financing methods pioneered by Enron in the 1990s—including derivatives, synthetic credit default swaps, and deals financed with little or no equity.[4]

Eventually, crony dealings led to Aubrey McClendon's resignation as CEO of Chesapeake. News emerged that he had been logging family vacation trips to Europe on one of the company jets as business travel, while Chesapeake employees were doing millions of dollars' worth of personal work at McClendon's home. He had put longtime friends on the Chesapeake board and showered them with compensation. He maintained personal stakes in company wells and used these as collateral for $1.55 billion in loans; at the same time, he was borrowing money from a company board member and running a private two hundred million dollar hedge fund from Chesapeake offices.[5]

Disgrace couldn't keep Aubrey McClendon down for long. Since his formal departure from Chesapeake Energy, he has formed a new company called American Energy Partners, with headquarters down the street from his old Chesapeake office in Oklahoma City.[6]

Clearly, not all oil and gas companies specializing in fracking mimic Chesapeake's extravagance and opacity. However, if the analysis in Chapter 3 is even approximately correct, then the shale industry is on shakier ground than many believe. Perhaps the fracking companies' business model simply reflects the problematic nature of the resources they pursue, and the price and investment structures needed to get those resources out of the ground and to market.

Chesapeake and other shale gas and tight oil companies have made extravagant claims about how communities, households, and the nation as a whole benefit economically from fracking. There's no question: a lot of money has changed hands as a result of shale gas and tight oil development during the past decade. Billions of dollars have been spent in drilling, and billions have been returned in sales of oil and gas. In this chapter, we'll try to answer the question: *Who really benefits?*

COMMUNITIES

The boomtown syndrome is as old as the petroleum industry itself. Once commercial deposits of oil or gas are confirmed, drillers and speculators arrive by the truckload, driving up prices for just about everything. Prostitution and motor traffic proliferate; peace and quiet disappear. Years later, after local citizens have spent their money from drilling leases, the oil or gas begins to peter out. High-paid workers leave town, and the local economy deflates.

This predictable syndrome tends to characterize fracking operations even more than conventional oil and gas development, because shale gas and tight oil per-well production rates tend to decline so steeply (making the boom briefer and the decline steeper), and also because damage to the environment and to local roads, public health, and community solidarity can be much more serious.

Here is a typical assessment of the economic boon from fracking, lifted from an industry-funded website:

> Hydraulic fracturing has . . . boosted local economies—generating royalty payments to property owners, providing tax revenues to the government and creating much-needed high-paying American jobs. Engineering and surveying, construction, hospitality, equipment manufacturing and environmental permitting are just some of the professions experiencing the positive ripple effects of increased oil and natural gas shale development."[7]

There's definitely some truth here. Consider the example of Bradford County, Pennsylvania. In recent years its economy had been in decline as manufacturing jobs moved to China. But now, as Chesapeake Energy and other operators are fracturing the Marcellus shale beneath the county to release billions of cubic feet of natural gas, the economy is flourishing. The county has retired five million dollars in debt and has lowered real estate taxes by 6%. A Bradford County Commissioner has called fracking "an economic game-changer for the entire area."[8]

However, studies that look at the bigger picture reach more nuanced conclusions. A report by the Keystone Center in Pennsylvania found that "the Marcellus Shale is making a small positive contribution to recent job growth in Pennsylvania."[9] New (and often temporary) jobs are being offset by damage to existing industries, including tourism and the Pennsylvania hardwoods industries. In Bradford County, cited above, the Pennsylvania Department of Environmental Protection has recorded upwards of six hundred environmental violations from fracking, and the consequences for farmers have been severe—including contaminated water, plummeting property values, and sickened livestock. Many farmers have simply given up in the face of these challenges. Meanwhile, Pennsylvania research has also found that many of the new jobs go to skilled out-of-state workers who fly in, drill, and fly home. The jobs for locals generated by fracking typically last for only about two to three years.[10]

All of this should be fairly predictable and unsurprising. Historically, regions that rely on resource extraction as an economic pillar often underperform when compared to other regions, especially when viewed over the long term. More wealth is typically created in places that *use* energy and minerals for manufacturing and trade than in ones where resources are mined. For example, coal areas in West Virginia continue to be pockets of poverty despite decades of mining activity. The long-term jobs created there often pay little, and other industries—including agriculture— are driven out by the ensuing environmental damage.[11]

In both Pennsylvania and New York, drilling companies are moving into the poorest counties first—and not just because that's where the shale resources are located. Economically struggling areas are often targeted because the locals are less likely to engage in anti-fracking activism. People who desperately need leasing money or temporary employment may be willing to overlook environmental damage, even if it impacts their own land and homes, and they can also be counted upon to take the industry's side when community disputes arise over air or water quality.

Meanwhile, people subsisting on fixed incomes (such as elderly rent-ers) who *don't* receive income from drilling leases or truck-driving jobs may have to move out because they can't afford soaring rents and food prices.

Local governments benefit temporarily from increased tax revenues during drilling booms. But costs to repair road damage—sometimes run-ning into the millions of dollars—may outweigh that short-term bonus. In 2012, the State of Texas received about $3.6 billion in severance taxes from all oil and gas produced in the state (from conventional wells as well as those in fracked shale). But during that same year, the Texas Department of Transportation estimated damage to Texas roads from drilling opera-tions at $4 billion. Arkansas has taken in roughly $182 million in severance taxes since 2009, but costs from road damage associated with drilling are estimated at $450 million. Roads designed to last 20 years are requiring major repairs after only 5 years due to the constant stream of overweight vehicles ferrying equipment and water to and from fracking sites.[12]

The influx of workers also puts pressure on schools and hospitals. Yet it makes little sense to expand these facilities permanently, given the temporary nature of drilling booms. Meanwhile, police have to deal with increased crime rates, including (in Colorado, for example) high rates of methamphetamine usage among drill crews.[13]

Disputes about fracking within communities can also strain the very process of democracy. When the city of Longmont, Colorado, enacted regulations to make residential neighborhoods, schoolyards, and the city's open spaces off-limits to drilling, Governor John Hickenlooper sued, con-tending that more lenient state regulations took precedence. In response, Longmont citizens launched an initiative to ban fracking altogether within the city. Though overwhelmingly outspent by industry money, the ban initiative carried by a remarkable 60/40 margin. Industry, backed by the state, has sued to overturn the ban.

As Deborah Rogers points out in her report, "Shale & Wall Street," the oil and gas companies "are not in business to steward the environment, save the family farm, or pull depressed areas out of economic decline. If these things should by chance happen, they are merely peripheral to the primary mission of the companies."[14] Yet the *promise* of benefits to com-munities helps these companies achieve their real primary goal—which is to extract hydrocarbons as cheaply and efficiently as possible and to sell them at the highest price that can be realized. Expectations of jobs and tax revenues can deter investigations into environmental and health problems, and delay regulations.

For communities that have endured environmental insults, human health impacts, and costs to road infrastructure, all for the sake of income from fuel production, it is perhaps the bitterest of ironies when oil and gas companies simply refuse to pay promised royalties. This is by no means standard operating procedure within the industry, but it does happen.[15] To mention just one example: in 2012 Chesapeake Energy paid five million dollars in settlement of a lawsuit brought by Dallas/Fort Worth Airport over significant underpayment for gas produced from horizontal wells beneath airport property.[16]

THE NATION

In their many opportunities to testify before Congress, oil and gas industry representatives have repeatedly painted a glowing picture of how America is benefiting from expanded shale gas and tight oil development. Here is Daniel Yergin speaking on the topic of shale gas in 2011:

> This abundance of natural gas is very different from what was expected a half decade ago. It was then anticipated that constraints on domestic natural gas production would result in high prices for consumers and the migration of gas-using industries—and the jobs that go with them—out of the United States to parts of the world with cheaper supplies. The United States was also expected to be importing substantial amounts of natural gas in the form of liquefied natural gas (LNG). That would have added as much as $100 billion to our trade deficit. None of that has occurred. Instead, the United States has become, except for imports from Canada, mostly self-sufficient. . . . Gas prices have fallen substantially, lowering the cost of gas-generated electricity and home heating bills. Several hundred thousand jobs have been created in the United States. Gas-consuming industries have invested billions of dollars in factories in the United States, something which they would not have expected to do half a decade ago—creating new jobs in the process. The development of

shale has created significant new revenue sources for states—for the state of Pennsylvania and localities in that state, for example, $1.1 billion in revenues in 2010.[17]

All true. But the implication is that with expanded drilling we can expect much more of the same. Not so true.

As for tight oil and its impact on America's petroleum production and imports, here is Yergin speaking to Congress in 2013:

Net imports of crude will continue to decline. . . . [W]e will see the Western Hemisphere, and North America in particular, moving towards greater self sufficiency. At the same time, the very large, technically advanced refining complex on the Gulf Coast—along with the shifting domestic product demand—will put the United States in the position to continue to expand exports of refined products. . . . [E]xpanded domestic supply will add to resilience to shocks and add to the security cushion. Moreover, prudent expansion of US energy exports will add an additional dimension to US influence in the world.[18]

The impression we are left with is that it's all good news in the oil and gas world, and America should be cheering. Yet that's far from being an accurate portrayal. What's most objectionable is the reassuring sense of permanence that characterizes Yergin's description of our national energy picture. Domestic supplies of oil and gas will increase . . . *but for how long?* A "prudent expansion" of US energy exports could only be based on the assumption that the current spate of growing production will continue for decades. Yet, as we have already seen, the EIA anticipates a peak in Bakken tight oil extraction in 2017. And if the analysis in Chapter 3 is accurate, US shale gas production will begin to taper off at roughly the same time.

The shale gas industry has touted natural gas exports as a way to improve the US trade balance. Indeed, despite the fact that the United States remains a net natural gas importer, gas export efforts are under way: Dominion Corporation will begin construction on an LNG export project at its Cove Point terminal in Maryland in 2014, with contracts for delivery to Japan and India; Cheniere Energy is converting its Sabine

Pass LNG import facility in Louisiana into an export terminal; and United LNG has signed agreements with India for the long-term supply of LNG by way of its offshore Main Pass Energy Hub, also in Louisiana.[19] Meanwhile, Congress is on board with LNG exports—presumably as a way to boost the nation's foreign revenues.[20]

But the industry's actual motives have nothing to do with improving America's balance of trade. Potential LNG importers in Japan, India, and China will pay upwards of $15 per million Btus for natural gas, while the American domestic price hovers around $4. The natural gas industry wants to export its product for one reason only: to get a better price. If American users want that same gas, they will have to pay more. LNG exports will drive up the US price of natural gas: it's simple economics, and no one who has seriously looked into the matter claims otherwise.

The natural gas industry is doing everything it can to substantially increase US natural gas prices, yet the same industry claims low gas prices as a benefit of its practices. The cognitive dissonance inherent in this situation is perhaps lost on most casual observers, but not on the American chemicals industry and electric power utilities—big users of gas. Recall Daniel Yergin's statement (quoted above) that "gas-consuming industries have invested billions of dollars in factories in the United States, something which they would not have expected to do half a decade ago—creating new jobs in the process." Are those jobs about to go away as gas prices shoot back up? Will power generators that switched from burning coal to natural gas in order to take advantage of low gas prices now switch back to coal?[21]

Industry boosters like Daniel Yergin often cite jobs as one of the principal benefits to America from fracking. Yet, as we have already seen, fracking's actual jobs record has been oversold. Industry-funded studies often include professions such as strippers and prostitutes in their totals of new jobs created.[22] Jobs for actual oil and gas industry workers have accounted for less than one-twentieth of 1% of the overall US labor market since 2003,[23] according to the US Bureau of Labor Statistics, and where those jobs relate to fracking, they will gradually disappear as plays are drilled out and production declines.

In a research note for *Capital Economics* (April 2013), Paul Dales argued that the oil and gas boom has provided only a modest economic boost to the US economy in recent years:

Since June 2009 the volume of oil and gas extraction has risen by 24%. Over the same period the production of mining machinery has risen by 47% and the output of mining support services, which includes oil and gas drilling, has leapt by 58%. . . . **But that rise explains only a small part of the economic recovery.** Admittedly, it is responsible for a fifth of the 18.3% increase in overall industrial production. Given that the oil- and gas-related sectors account for only 2.5% of GDP, **they have contributed just 0.6 percentage points (ppts) to the 7.6% rise in GDP.**[24] [Emphasis in original]

Perhaps the biggest real impact of fracking on the nation is at the macro scale of energy policy. As a result not just of temporarily increased production, but also of exaggerations from the industry (and spokespeople like Daniel Yergin), the United States is failing to plan for a future in which hydrocarbons are more scarce and expensive; failing to invest sufficiently in renewable sources of energy, and in low-energy infrastructure such as electric light rail; and failing generally to do what every nation must in order to survive in a century of rapidly destabilizing climate—which is to *reduce dependency on fossil fuels as quickly as possible.* US politicians of every stripe end up adopting the attitude, "Yes, of course we should be reducing fossil fuel consumption in order to avert the worst climate change impacts—but with the prospect of energy independence, jobs, and economic growth all flowing from shale gas and tight oil, how could we possibly say no?"

Yet, as we have just seen, these promises are largely unrealistic. Meanwhile, the costs of continued oil and gas dependency, which will be borne mostly in the future, already constitute an invisible though mountainous burden, compounding daily.

Throughout, fracking boosters appear to maintain a Marie Antoinette-like attitude toward the American people, saying, in effect, *"Let them eat hype."*

THE OIL AND GAS INDUSTRY

When we inquire who benefits from the fracking frenzy, the intuitively obvious answer is, *"the oil and gas industry, of course."* Yet this may be a simplistic assumption.

Drilling companies have seen increased revenues as a result of shale gas and tight oil production. This is primarily true for the outfits that managed to lease land in the core areas of shale plays, where wells are more productive and profitable.

The industry as a whole has scored big in terms of public relations: the promise of plenty has largely shifted the conversation in America from worry about high energy prices, supply vulnerability, and climate risks, to ebullience over the illusion of "energy independence."

However, as we have just seen, many shale gas (and some tight oil) operators in non-core areas have been losing money on production. Only a few companies are profiting handsomely. Clever operators can squeeze profits out of money-losing field operations with "pump and dump" stock schemes, selling stocks to gullible pension fund managers, or selling bundles of drilling leases (of highly variable quality) to foreign investors. A few early birds (such as Aubrey McClendon) have gotten spectacularly rich, while building financial structures packed with risk.

One of these risks consists of price volatility. Shale gas and tight oil production only makes economic sense when fuel prices are high. But (as gas drillers have learned, to their bitter disappointment) there is no guarantee that prices will stay high enough, long enough, for investments to pan out. If the US economy were to fall back into recession, energy demand would drop and so would oil and gas prices. Drillers would have no choice but to idle their rigs and accept losses.

Perhaps that's what led Aubrey McClendon to say in an investor call in 2008, "I can assure you that buying leases for x and selling them for 5x or 10x is a lot more profitable than trying to produce gas at $5 or $6 mcf."[25] Gas prices wound up dropping to less than $2 per thousand cubic feet (mcf).

Another risk comes from potential liability for environmental and human health damage. Producers try to contain risk with ongoing improvements to their procedures, but also with nondisclosure agreements. States and counties are increasingly restricting and even banning fracking out of concern for human and environmental health, and the result could be a sharp decline in potential revenues for operators—and therefore a drop in stock value and an increase in borrowing costs. A general lack of transparency in the industry makes it difficult to ascertain the total payments made to date for settlement of damage claims. But worries about declining water and air quality pose a growing public relations hurdle for the companies.

Still another risk comes from the very hype that drives investment toward the industry. When projects go sour, investors flee, and suspicion grows among the investment community that shale production is merely a bubble. This perception is bolstered by companies pulling out of shale plays or deciding against funding, for example, a pipeline to North Dakota—presumably out of recognition that volumes of crude produced will not be high enough, or last long enough, to make such a pipeline pay off.[26] It's as though drillers are admitting they don't believe the hype themselves.

Fracking is helping the industry as a whole bring more product to market. But the top international oil companies—ExxonMobil, Shell, BP, Chevron, and Total—have seen their overall production decline by over 25% since 2004.[27] Meanwhile capital spending by the industry has nearly doubled. The only factor forestalling economic bloodshed for the majors is high oil prices: while they must invest more upstream to produce fewer barrels, each barrel sold now brings in more revenue.[28]

For a few players, the shale boom is a bonanza. Yet overall, despite protests to the contrary, evidence suggests the oil industry has entered its sunset years.

WALL STREET

If, in the final analysis, the nation as a whole and the impacted communities within it lose more from fracking than they gain, and the oil industry is seeing diminishing returns on its burgeoning investments, then who *does* stand to benefit?

In the Introduction and Chapters 2 and 3, we noted how, in many instances, gas prices have been driven down to a level below industry's production cost. Low prices have in turn been cited as economic benefits of shale development. Yet aside from having gained a PR talking point, the industry itself has actually been hurt by low prices. Chesapeake Energy has not only reduced drilling, but sold off hundreds of millions of dollars' worth of assets to cover unsustainable debt loads. BP has been forced to write off nearly two billion dollars in assets. Rex Tillerson, the

CEO of ExxonMobil, told the Council on Foreign Relations in New York City in June 2012, "We're losing our shirts [on shale gas production]. We're making no money. It's all in the red."[29]

Why did the industry shoot itself in the foot by overproducing shale gas? In some respects, the glut resulted inevitably from a land rush that occurred in the early years of the shale boom, when companies were leasing as much land as possible, as quickly as possible: the terms of drilling leases required drilling sooner rather than later, even if that meant oversupplying the market. But this is not a sufficient explanation for the price plunge. For a deeper understanding of the industry's puzzlingly self-destructive behavior, we must follow the money.

In a *New York Times* investigative article ("After the Boom in Natural Gas," October 20, 2012), Clifford Krauss and Eric Lipton wrote, "Like the recent credit bubble, the boom and bust in gas were driven in large part by tens of billions of dollars in creative financing engineered by investment banks like Goldman Sachs, Barclays and Jefferies & Company." The article details how this "creative financing" forced drillers to keep drilling even when each new well represented a financial loss.[30]

Deborah Rogers, a former Wall Street financial consultant and member of the Advisory Council for the Federal Reserve Bank of Dallas from 2008 to 2011, has further traced the toxic connections between major investment banks and shale gas/tight oil operators in her report, "Shale and Wall Street: Was the Decline in Natural Gas Prices Orchestrated?"[31] She writes:

> In order for a publicly traded oil and gas company to grow extensively, it must manage not only its core business but also the relationship it enjoys with its investment bankers. Thus, publicly traded oil and gas companies have essentially two sets of economics. There is what may be called field economics, which addresses the basic day-to-day operations of the company and what is actually occurring out in the field with regard to well costs, production history, etc.; the other set is Wall Street or "Street" economics. This entails keeping a company attractive to financial analysts and investors so that the share price moves up and access to the capital markets is assured. (p. 6)

It was "street economics" rather than "field economics" that drove the gas glut and price rout, according to Rogers, who notes that the price decline "opened the door for significant transactional deals worth billions of dollars and thereby secured further large fees for the investment banks involved. In fact, shales became one of the largest profit centers within these banks in their energy M&A portfolios since 2010." (p. 1) She concludes that the glut was engineered in large measure "in order to meet financial analysts' production targets and to provide cash flow to support operators' imprudent leverage positions." (p. 1) When natural gas prices tanked,

> Wall Street began executing deals to spin assets of troubled shale companies off to larger players in the industry. Such deals deteriorated only months later, resulting in massive write-downs in shale assets. In addition, the banks were instrumental in crafting convoluted financial products such as VPP's (volumetric production payments); and despite the obvious lack of sophisticated knowledge by many . . . investors about the intricacies and risks of shale production, these products were subsequently sold to investors such as pension funds. Further, leases were bundled and flipped on unproved shale fields in much the same way as mortgage-backed securities had been bundled and sold on questionable underlying mortgage assets prior to the economic downturn of 2007. (p. 1)

Wall Street benefits from manias—at least in the short term. Investment banks make money on sales of shares in companies whose activities spur speculative bubbles. They also profit from mergers and acquisitions when bubbles burst and companies go bust. For the most part, it's not their money being invested—it's more likely yours, if you have any kind of retirement account.

There are a lot of people benefiting from the shale gas and tight oil boom, ranging from drillers to landowners to hoteliers, but arguably none have profited more than investment bankers. And when oil and gas production falls and the fortunes of drillers, landowners, and hoteliers plummet along with it, the bankers will most likely continue to do just fine, thanks.

SNAKE BITES

1 **THE ENERGY PUNDITS SAY:**
We will never run out of fossil fuels!

THE REALITY IS:
True—but only because vast quantities of fossil fuels are neither *economically* nor *technically recoverable*.

2 **THE INDUSTRY SHILLS SAY:**
There are immense supplies of methane hydrates, oil shales, and other unconventional sources that we only need the right technology to exploit.

THE REALITY IS:
The "right technology" can also transport humans with light sabers to the time before dinosaurs. Studies that examine the energy return on energy invested (EROEI) for these unconventional energy sources have not been remotely promising.

*Hydrocarbons are so abundant that if we burn a substantial portion of them, we risk a climate catastrophe beyond imagining. Even so, there aren't enough **economically accessible, high-quality** hydrocarbons to maintain world economic growth for much longer.*

Chapter Six

ENERGY REALITY

D uring the past year, article after article in the mainstream press has gushed over the prospects for American oil independence and natural gas exports, while ignoring the context—an ever-increasing requirement for the investment of capital and energy in the extraction of fast-depleting and often poorer quality fuels.

The media's euphoria was perhaps epitomized by Charles C. Mann's lead article in the May 2013 issue of *Atlantic* titled "What If We Never Run Out of Oil?" The magazine's cover proclaimed, in tall capital letters, "WE WILL NEVER RUN OUT OF OIL"—which of course is true: the Earth's crust will always contain immense amounts of crude. It's just that we won't be able to afford to extract most of it because doing so would take either too much money, too much energy, or both. Continuing the theme, the article's subtitle asked a startling question—*"What if fossil fuels are not finite?"*—which implies uncertainty as to whether the Earth is a bounded sphere or a plain extending endlessly in four directions. Title, subtitle, and headline were presumably intended as attention-grabbers: the article itself was serious and thoughtful—though, as I hope to show, profoundly misleading.

In this chapter, we will first address a few of Charles Mann's claims in the *Atlantic* article and then proceed to the much more important discussion of our real energy prospects.

OTHER UNCONVENTIONAL HYDROCARBONS

As a warm-up for touting America's shale gas and tight oil prospects, Mann spent the opening pages of his article introducing readers to the truly gargantuan potential of methane hydrates—a frozen hydrocarbon resource locked in seabeds and Arctic tundra. "Estimates of the global supply of methane hydrates," wrote Mann, "range from the equivalent of 100 times more than America's current annual energy consumption to 3 million times more." Numbers that big numb the brain.

This should have been the appropriate point in the article to explain the resource pyramid, and to inform readers that nearly all (if not all) of the world's methane hydrate resources rest at the bottom of the pyramid, where they are economically inaccessible and likely to remain so. Japan has conducted the world's most extensive research on commercial extraction of this resource, and, as Chris Nelder noted in a rebuttal to Mann's piece, "Japan's experiment so far has taken 10 years and $700 million to produce four million cubic feet of gas, which is worth . . . about $50,000 at today's prices for imported LNG in Japan."[1] Mann, in a reply to Nelder's rebuttal, countered that technology R&D costs should not be taken into account in assessing the commercial viability of a resource.[2] That's arguable. However, using the word *supply* to describe these resources, as Mann does ("estimates of the global *supply* of methane hydrates"), is clearly misleading, because virtually no hydrates are actually being supplied.

Whatever the size of the resource base, *economic reserves* of methane hydrates are currently roughly zero. We simply do not know if the extraction of these resources can ever be accomplished at an energy or economic profit. Most of the geologists I've spoken with on the subject are highly skeptical. The EROEI (energy return on energy invested) for commercial methane hydrate extraction is unknown, but preliminary indications are not encouraging. A study of the EROEI for electrical heating of methane hydrate deposits located at depths between 1000 and 1500 meters yielded ratios from less than 2:1 up to 5:1, depending on the source of electricity. The authors of the study emphasize that this is only one of the energy inputs that must be taken into account.[3]

Other authors mimic Mann's hydrocarbon hyper-enthusiasm when discussing "oil shale" (more properly termed *kerogen*), which is extraordinarily abundant in Colorado and Utah. The United States has the largest deposits of this resource in the world, amounting to nearly 4.3 trillion barrels of oil equivalent. Novice commentators often take that number, divide it by America's annual oil consumption (roughly 7 billion barrels), and arrive at the mind-melting conclusion that the nation is sitting on six hundred years' worth of oil.

But *kerogen is not oil*. It is better thought of as an oil precursor that was insufficiently cooked by geological processes. If we want to turn it into oil, we have to finish the process that nature started; that involves heating the kerogen to a high temperature for a long time. And that in turn takes energy—lots of it, whether supplied by hydroelectricity, nuclear power plants, natural gas, or the kerogen itself. Therefore, the EROEI in extracting and processing oil shale is bound to be pitifully low. According to the best study to date, by Cutler Cleveland and Peter O'Connor, the EROEI for oil shale production would be about 2:1.[4] That tells us that oil from kerogen will be far more expensive than regular crude oil—right up until the time when regular crude oil itself becomes uneconomic to produce.

In "Drill, Baby, Drill," after carefully analyzing US shale gas and tight oil prospects, David Hughes proceeds to assess global unconventional oil resources; in the pages devoted to oil shale he points out:

> [W]ith oil shale, as with all hydrocarbon accumulations, there are variations in quality between basins and there are sweet spots within basins. For this reason, the relatively high quality oil shale resources within the Piceance Basin have received the most attention in recent years with pilot projects conducted by the oil majors Shell, Chevron, and ExxonMobil, as well as a number of smaller companies. None of these pilots has resulted in commercial scale production and Chevron has recently abandoned its operations. [p. 123]

Again: the resources are immense, yet economic reserves are minuscule to nonexistent. Sometimes this can be hard to explain to the layperson. I recall all too many instances where I have carefully described to a lecture audience how *it takes energy to get energy,* pointing out that the energy profit from the production of kerogen resources is abysmal—only to hear an audience member insist that there must be some dark conspiracy preventing America from exploiting these unfathomable energy riches.

Other unconventional hydrocarbons are viable, but still problematic and often overestimated. Canada's tar sands (better termed *bitumen*) are clearly an economic source of fuel, and again the resource is immense—1.84 trillion barrels, or about 60 years of global oil consumption at current rates. But only about a tenth of that resource is currently counted as reserves. The EROEI for tar sands production is poor, between 3:1 and 6:1 by most estimates. "Syncrude"—synthetic crude oil made from bitumen—is profitable to produce only because oil prices are high and natural gas prices are low. (Heat from gas is often used to liquefy the tar.)[5] Bitumen, like all nonrenewable resources, is subject to the low-hanging fruit extraction principle: the very best resources are being mined first—which means that, as time goes on, the requirement for financial and energy investment per barrel of finished syncrude will tend to increase.

Like tar sands, Arctic and deepwater sources of oil are currently economic—at least in some instances. For the United States, the Gulf of Mexico is the site of nearly all current deepwater production. The Gulf boasts a total of almost 70 billion barrels of reserves plus estimated "undiscovered technically recoverable resources." Shell is currently developing some of the deepest wells ever drilled, in nearly two miles of ocean water two hundred miles south of New Orleans.[6] Deepwater petroleum resources also exist off America's Pacific and Atlantic coasts, and the North Slope of Alaska. In all instances, deepwater drilling entails high environmental risks (recall the Deepwater Horizon disaster in the Gulf of Mexico in 2010), but especially so in ice-choked Arctic waters. The realistic prospect is for a combined production rate no greater than 1.7 million barrels per day from all US deepwater projects through 2035 (which equates to about 2% of world crude oil consumption), after which production will decline.[7] Deepwater projects typically suffer from high production costs: a single well may require the investment of $100 million or more. Drilling costs are highest in the Arctic, as Shell recently discovered: in January 2013, a Shell drilling rig called the Kulluk broke free from a tow ship in stormy seas and ran aground near the island of Kodiak. The immediate loss was assessed at $90 million, and there were no oil production revenues from the project to offset it.[8] Deepwater exploration and production are only profitable when oil prices are high.

Our problem is not that there aren't enough hydrocarbon molecules in the ground. (Charles Mann is right on that point.) There are certainly plenty to fry the planet many times over, if we were to burn them all. Instead, our most pressing energy conundrum—from a purely economic standpoint—is declining EROEI. We built industrial societies on high-EROEI fuels that enabled a small amount of investment, and relatively few workers, to supply enough cheap, concentrated energy so that the great majority of citizens could use ever-increasing amounts of energy and thereby become more productive. Millions of farmers (who are traditional societies' primary energy producers) were freed up to become factory workers, salespeople, computer technicians, perhaps even hedge fund managers or journalists. During this time, labor productivity soared—not because people were working longer and harder, but because they were using more energy at their jobs (by way of machinery) to generate more wealth. Urbanization, globalization, specialization, rapid economic growth—none of these would have been possible without increasing flows of energy that was spectacularly cheap in both monetary and energy terms.

Lower the overall EROEI of the energy system of a modern industrial society and the predictable result is a requirement for more investment in the energy sector and for more workers there as well. Economic growth slows, stalls, or reverses; jobs in non-energy sectors disappear; globalization falters. Meanwhile, more expensive energy translates to a stagnation or even decline in worker productivity.

This is exactly what we are beginning to see.

GEOLOGY VERSUS TECHNOLOGY

A key point of Charles Mann's article in *Atlantic* was that technology changes the game. It was new technology (hydrofracturing and horizontal drilling) that made a torrent of new shale gas and tight oil production possible. Technology is driving expanded extraction of Canada's tar sands. Technology could make methane hydrates accessible and could make Arctic oil easier to reach. We simply don't know how much of the world's currently inaccessible, vast, unconventional hydrocarbon resource base can be turned into economic reserves through further advances in technology. Therefore (so goes the argument), to discount the likelihood of a future of cheap, plentiful fossil fuels simply because we're depleting reserves of conventional fuels is foolish.

A discussion about the unknown capabilities of future technology could easily descend into the trading of empty claims based on contrasting prejudices. We can avoid that wasted effort by clarifying the essence of the dispute and then examining the evidence. The question we really need to answer is this: *Can technology improve the overall EROEI of fossil fuel extraction enough to overcome declines in resource quality resulting from the depletion of conventional fuels?* Now, let's look at the relevant facts.

Technology can certainly improve the EROEI of oil, gas, and coal production. Examples of energy-saving innovations include clustered pad drilling for shale gas, cogeneration in tar sands production, longwall coal mining, and closer well spacing in tight oil plays.[9] The industry is always looking for ways to save money, and efficiency measures undertaken in order to reduce investment requirements usually end up saving energy as well.

Technology or geology: Which horse will win? In the end, geology is destined to triumph. Energy efficiency moves us in the direction of solving the EROEI dilemma, but it is always subject to the law of diminishing returns: the first 5% increase in energy efficiency typically costs less than the next 5%, and so on. Meanwhile, the effects of depletion compound: fossil fuels *are* finite, regardless of any attention-grabbing headline to the contrary, and the extraction costs for fossil fuels tend to rise exponentially as resource quality declines below certain thresholds. Efficiency improvements will eventually be overcome by the sheer physical burden of harvesting hydrocarbons that are increasingly deep and dispersed.

However, "in the end" and "eventually" are too vague to be helpful. What we really need to know about is the short term—say, the next 20 years. During the next two decades, society will still be largely dependent on fossil fuels even if it makes a substantial effort to reduce that dependency in order to avert catastrophic climate change. In fact, fossil fuels will be doing double duty: they will be keeping major sectors of our current economy going (most essentially, our transport and food systems), while also providing energy for the manufacture of millions of solar panels and wind turbines (it's only just this year that the world's solar power plants installed to date have produced as much energy as was required to build them).[10] Whatever fossil fuels we continue to use will have to be highly productive—especially since most renewables have energy profit ratios lower than those of fossil fuels (as we will see in the next section). To focus the relevant question even further: Can improvements in extraction technology enable fossil fuels to keep modern, complex societies economically viable during this crucial transition period?

The signs are not favorable. Currently, the overall EROEI for fossil-dominated global energy is declining. That's the conclusion of a boatload of ongoing research by a growing number of scientists.[11] Charles Hall is the father of EROI research. (He prefers the term *EROI*, or *energy return on investment*, because it considers capital and environmental investments as well as energy investments in energy production; EROEI refers only to the investment of energy in energy production.) He writes that "the world's most important fuels, oil and gas, have declining EROI values. As oil and gas provide roughly 60 to 65% of the world's energy, this will likely have enormous economic consequences for many national economies."[12] Hall's finding is based on numerous recent studies: "This pattern of declining EROI," he writes, "was found for US oil and gas (Guilford et al.), Norwegian oil and gas (Grandell et al.), Chinese oil (Yan et al.), California oil (Brandt), Gulf of Mexico oil and gas (Day and Moerschbaecher), Pennsylvania gas (Sell et al.), and Canadian gas (Freese)."[13]

A few energy financial analysts have explored the implications of EROEI, often without observing Hall's methodological rigor and without properly citing his original work in this field. Andrew Lees of UBS, writing in *The Gathering Storm*, has argued that global EROEI is currently about 20:1, deriving this figure from energy's 4 to 5% share of world GDP. Given recent trends, Lees calculated that the ratio might fall to 5:1 over the next decade, which would translate to a massive disruption of the world economy.[14] Discussing Lees's conclusions, the *Economist* magazine mused that "the direction of change seems clear. If the world were a giant company, its return on capital would be falling."[15]

Tim Morgan, of the London-based brokerage Tullett Prebon (whose customers consist primarily of investment banks), discussed the averaged EROEI of global energy sources in a recent *Strategy Insights* report, noting:

> [O]ur calculated EROEIs both for 1990 (40:1) and 2010 (17:1) are reasonably close to the numbers cited for those years by Andrew Lees.

For 2020, our projected EROEI (of 11.5:1) is not as catastrophic as 5:1, but would nevertheless mean that the share of GDP absorbed by energy costs would have escalated to about 9.6% from around 6.7% today. Our projections further suggest that energy costs could absorb almost 15% of GDP (at an EROEI of 7.7:1) by 2030. Though our forecasts and those of Mr. Lees may differ in detail, the essential conclusion is the same. It is that the economy, as we have known it for more than two centuries, will cease to be viable at some point within the next ten or so years unless, of course, some way is found to reverse the trend.[16]

In two of three primary fossil fuel energy sectors, extraction costs are rising. Technology may be winning a battle here or there, but the evidence shows that, as of now, it's losing the war.

Charles Mann discusses EROEI briefly in his *Atlantic* article, pointing out one apparently bright spot in the landscape: shale gas. He reports that the EROEI of shale gas is a shining 87:1. He doesn't provide a source for this figure, but it apparently comes from a study by Bryan Sell, David Murphy, and Charles Hall.[17] In it, the authors analyzed "tight gas" production in Indiana County, Pennsylvania, using drilling and production data from before 2003. In other words, the data do not reflect the energy costs associated with the new and more complex and costly technology associated with horizontal drilling and hydrofracturing. The authors did not attempt to account for transmission and processing energy costs (which might lower the result by up to half). They were also very conservative in accounting for other energy costs. Crucially, they note that "highly complex-drilling environments, such as some shale gas reservoirs, could ultimately show relatively low EROI values." No evidence suggests that the technology of fracking has actually *raised* the EROEI for natural gas production. (It temporarily lowered prices, but only by glutting the market.) Moreover, in their concluding remarks, Sell, Murphy, and Hall discuss the spectacularly high decline rates of shale gas wells and note, "catastrophic drops in gas supply can be expected if shale gas is relied upon as a replacement [for] conventional gas."

There is good reason to think that the EROEI of shale gas is probably higher in the Marcellus (where operators are still drilling in "sweet spots") than in older plays like the Barnett, Haynesville, and Fayetteville.

As core areas are drilled out and rapidly deplete so that drillers are forced to move to areas with lower productivity, the overall energy return for shale gas drilling and production is probably declining rapidly. Indeed, David Hughes, in "Drill, Baby, Drill," speculates that if all energy inputs are properly accounted for, the EROEI of shale gas in the older plays may be 5:1 or less on average.[18]

Technology can trump geology for a while, at least in certain instances.[19] But we have entered a new era in which geology is negotiating harder all the time, and the costs of new technology often outweigh the economic benefit promised. Some fossil fuels (coal and gas) still have a relatively high EROEI, but oil is crucial to the global energy mix since it fuels virtually all transport, and oil's energy profit ratio is plummeting.[20] The economy is not likely to respond in steady increments to declines in energy profitability. This is how Tim Morgan at Tullett Prebon puts the matter: "[T]he critical relationship between energy production and the energy cost of extraction is now deteriorating so rapidly that the economy as we have known it for more than two centuries is beginning to unravel."[21] Failing to notice this historic shift, while celebrating a temporary breakout in oil and gas production numbers in Texas, Pennsylvania, and North Dakota, seriously hampers our ability to adapt to dramatically and quickly changing circumstances.

RENEWABLE ENERGY

We need much more renewable energy, and we need it fast. We must replace fossil fuels in order to prevent a climate catastrophe. And we must leave oil, gas, and coal behind because they are depleting, nonrenewable fuels that will inevitably become more expensive and dirtier the longer we rely on them.

The EROEI for most renewables is lower than the historic energy profit ratios for fossil fuels (see Figure 13 in Chapter 1). But the EROEI of oil and gas is declining, while the EROEI of wind and solar photovoltaics is improving.[22] As we've just seen, the efficiency improvements in the production of fossil fuels are temporary, because they are quickly overcome by declining resource quality. But with renewable energy sources, technological improvements do not face the same headwinds. This is a crucial trend in our favor, and we should make the most of it.

Still, there are real hurdles to overcome.

The world's largest current renewable source of energy is hydroelectric power. It can't grow by very much, and building dams often creates enormous environmental problems.

The main renewable energy sources that *are* capable of significant growth are solar and wind. Both are intermittent; this can create challenges for grid operators. Typically those challenges are addressed by building energy storage capacity and by managing the grid to take advantage of a diverse portfolio of wind and solar generators sited in different places with different weather conditions. Some recent studies suggest that clever electricity supply and demand management could enable renewables to provide most, or perhaps even all, of America's power without serious difficulties.[23] However, the grid operator in Germany—a country with extensive experience in solar and wind—reports that, with high grid penetration, intermittency leads to problems like blackouts and brownouts, which in turn can damage electronic devices.[24]

For the world as a whole, growth in supply of renewable energy is not occurring at a sufficient rate to entirely displace fossil fuels any time soon. Some countries are seeing relatively quick adoption: Germany generated 23% of its electric power from renewables in 2012 (the proportion doubled in six years); Denmark achieved 41% renewable power; and Portugal, 45%. Here in the United States, Texas produced nearly 30% of its power from wind on some days last year. Yet the IEA notes that "worldwide renewable electricity generation since 1990 grew an average of 2.8% per year, which is less than the 3% growth seen for total electricity generation."[25] Moreover, there has recently been some slowing in the furious growth pace of solar installations in many countries because of reductions in government incentives, due in turn to the debt crisis in Europe and the squeeze on all older industrial economies from high oil prices.

Renewable energy boosters hope that falling prices will make solar and wind cheaper than fossil fuels, so that incentives will no longer be needed, and the growth rate for renewables will soar. Prices for both solar and wind have dropped steadily in recent years, and in some cases are competitive with natural gas (especially given the cost to utilities of hedging against gas price volatility). However, for the solar industry, low prices are a mixed blessing. Many photovoltaic (PV) producers are losing money, and factories are closing. A massive consolidation of the solar industry is under way.[26] Prices may have to rise in order for solar manufacturers to remain profitable. If that happens, the growth rate for solar penetration into electricity markets will be further constrained.

Another hurdle is the fact that solar and wind produce electricity, while transport runs on oil. How can we make transport energy renewable? All routes to that goal are problematic.

Electric vehicles offer a partial solution, but market penetration is not occurring nearly fast enough. And there are problems with high energy and materials costs for manufacturing batteries. There are no electric airliners on the drawing boards and probably never will be.

Hydrogen-powered vehicles have been hailed as a vector for renewably energized transport, but these have been very slow to deploy because fuel cells are expensive, and hydrogen is hard to store.

Advanced biofuels are another proposed solution. Companies are working to develop biofuels from city sewage, from contaminated grains and nuts, from cannery wastes and animal manures, and from forest wastes. Efforts are also under way to make liquid fuels from algae. Add up all these potential sources and they could nearly equal current transport energy; the remainder could be dealt with through better vehicle efficiency. But all biofuels have a low EROEI. Indeed, many of these potential fuel sources are likely to have a zero or negative net energy balance. Some make sense as ways of dealing with waste products, but as ways to economically produce energy—not so much. Alan Shaw, the chemist and former chief executive officer of Codexis, the first advanced biofuel technology company to trade on a US exchange, now says, "Cellulosic fuels and chemicals are not widely manufactured at commercial scale because their unit production economics have not yet been shown to be competitive with incumbent petroleum."[27]

EROEI is not the only criterion by which we should assess energy sources. We also need to take into account their environmental risks and their long-term viability. On these latter criteria, renewable energy sources score better than fossil fuels, though renewables do entail environmental costs (building solar panels and wind turbines requires extraction of depleting nonrenewable resources and generates pollution). However, without a high EROEI, renewable energy sources will never power the kind of growing, fast-paced consumer society that policy makers mistakenly believe to be the necessary goal of all economies.

Mark Jacobson at Stanford University and Amory Lovins of Rocky Mountain Institute say we can power the world entirely with renewable energy sources in 20 to 40 years with no real economic sacrifice.[28] Skeptics like Ted Trainer at the University of New South Wales say the transition will be expensive and littered with engineering nightmares.[29] One can cherry-pick data to support either position.

One thing we can say for sure: by the end of this century the world economy will be running mostly, if not entirely, on renewable energy sources, whether that economy is robust or withered, and whether or not we have made substantial investments in alternative energy. Fossil fuels simply won't get us to that far shore. Even if we don't know exactly what kind of ride they will give, renewables are the only boats we have that don't leak.

ENERGY SCENARIOS

It is, of course, impossible to predict exactly what our energy future will be, but current trends suggest a few likely possibilities.

Let's start with prospects for oil. During the past couple of years, global prices have bounced around within a band ranging from $95 to $115. This results from an uneasy supply balance maintained by the ongoing depletion of conventional oil fields and the simultaneous appearance of more expensive oil from unconventional sources. This is an inherently unstable dynamic. One might think that higher oil prices would inevitably follow as drillers are forced to move to ever-more expensive prospects, but this is not necessarily the case. A renewed global recession would cut energy demand; in that case, oil prices could fall significantly. With a dramatic reduction in trade and higher unemployment, we would also see declining overall oil production.[30] If the price of oil falls below $90 per barrel, new deepwater drilling will slow. At $80, new tar sands projects will be put on hold. At $70, nearly all drilling will be called off (except where required in order to maintain lease agreements). At $60, tar sands production from some existing projects will be throttled down.[31]

With cheaper oil, the economy might rebound somewhat; but then demand for oil would likely pick up again and so would prices. Altogether, the picture is bleak for oil economics.

The prospects for natural gas are not much better. Two trends are likely to drive gas prices higher. Currently, US drilling rates are down, so production will inevitably start to slide in the next couple of years as a result of the high per-well decline rates of shale plays and the drilling out of the core regions within each play. Also, if and when the United States begins exporting LNG, this will serve to push up domestic gas prices. These are not mutually exclusive developments, and if both happen, America could be facing both lower natural gas production rates and much higher prices.

There is one scenario in which natural gas prices would fall, but it's not a pretty one: it entails a serious economic recession that would destroy demand for the fuel through massive unemployment and a collapse of manufacturing.

Higher natural gas prices would be welcomed by the US coal industry, which has been struggling for the last few years under the onslaught of (temporarily) cheap shale gas. American coal producers want to export their product to China, which has nearly maximized domestic mining capabilities. China's options for new energy sources to fuel economic growth include imported oil, imported coal, imported LNG, nuclear, solar, wind, and shale gas; all are more expensive than the country's own fast-depleting coal. If China begins importing coal from the United States at the same time as American domestic natural gas prices soar (which would entice utilities to burn more coal once again), the result could be a spike in domestic coal prices. Higher coal prices would be abated only by a serious recession or falling gas prices. Several recent studies conclude that world coal extraction rates have little headroom: despite the vastness of the resource base, most of the high-quality, easily accessible coal is already gone.[32]

Altogether, fossil fuel prices appear to be on the verge of increased volatility: we will likely see more frequent and severe booms and busts within the oil, gas, and coal sectors. At the same time, accounting properly for energy costs in energy production, we will probably see less net energy delivered to society. And fossil energy will be generally less affordable. The overall EROI of society—the energy return on *all* investments in energy production, including financial as well as energy investments—will fall.

We have not discussed nuclear power thus far, and readers who see nuclear as a major future energy source will have found this frustrating.

However, I generally agree with the analysis of the *Economist* magazine, which recently published a special report calling nuclear power "The Dream that Failed."[33] Nuclear is just too expensive and risky. It was a technology that seemed to make sense in an earlier era of high fossil energy returns from minor investments, when enormous research, development, and construction costs for fission power could easily be shouldered. Today it is far more difficult to divert capital away from other energy projects. Even though nuclear electricity is inexpensive once power plants are built, the initial investments—several billion dollars per project, with inevitable cost overruns and the requirement for government loan guarantees and insurance subsidies—are now just too high a barrier. Currently, the industry is expanding in only a few nations, principally China—a country that gets most of its energy from cheap, high EROEI coal.

The only regions relatively immune to the economic whipsaw of fossil fuel dependency will be those reliant on renewable energy. But new investments in renewables, as we saw in the previous section, have slowed due to the systemic anemia of the Western economies and the false expectation of cheap and abundant natural gas for decades to come. The trend to ease back on renewable energy incentives cannot be allowed to continue. The world may have a fairly brief window of time in which major investments in renewable energy are feasible. Beyond that point, the volatility of fossil fuel prices and declining overall societal EROI may drain the vitality of economies to the point that financing major new projects will become ever more difficult. This is perhaps the most important reason that the conventional wisdom of a new golden era of oil and gas abundance must be countered.

In the worst case, societies may enter an ongoing maintenance crisis, seeking merely to keep basic services available as energy and capital contract in a self-reinforcing feedback loop.[34] The better-case scenario would start with major immediate investments in renewable energy. How it would unfold from there requires considerable speculation. Society would almost certainly need to adapt to economic stasis or contraction as a result of declining mobility and EROEI. It would also need to rebuild transport and food systems to use less overall energy and different energy sources. In the best-case scenario, we will tomorrow discover a new, abundant, cheap, high-EROEI energy source with no carbon emissions.[35] Betting on that highly unlikely event seems foolish; in all likelihood, we will have to settle for solar and wind. But we won't have even those if we don't start building panels and turbines at a ferocious pace.

A MIRAGE DISTRACTS US FROM HYDROCARBON REHAB

I have devoted a portion of this chapter to countering assertions in Charles Mann's *Atlantic* article not because he deserves scorn. Mann is no fossil fuel industry shill; he is a respected historian and the author of several excellent books (including *1491: New Revelations of the Americas Before Columbus*). He doesn't exaggerate the world's hydrocarbon prospects because he wants us to burn all that oil, gas, and methane hydrate. Quite the contrary; he is deeply concerned about climate change. The full sub-title to his article is "New technology and a little-known energy source suggest that fossil fuels may not be finite. This would be a miracle—and a nightmare." I chose Mann as a foil because he epitomizes the general failure of America's intellectual class to comprehend and communicate the complexity of our energy-economy-climate situation. It is an understand-able failure, but it may be a fateful one.

Perhaps the most concise way to convey this complexity is by way of two equally true statements:

- Hydrocarbons are so abundant that, if we burn a substantial portion of them, we risk a climate catastrophe beyond imagining.
- There aren't enough economically accessible, high-quality hydrocar-bons to maintain world economic growth for much longer.

Here is a public relations nightmare: how to convey these seemingly con-tradictory messages to people without confusing the bejesus out of them. How can concepts like "energy return on energy invested" be explained to an audience that barely understands what energy is? How can millions of half-somnolent television addicts be guided in understanding "fugitive methane emissions," "energy density," and a dozen other essential terms and concepts? Where are the cover stories in chattering-class magazines, the hour-long NPR interviews, the TV newsmagazine in-depth inves-tigative reports, and the congressional inquiries that explore the true intricacy and peril of our energy-economy-climate conundrum? Don't hold your breath waiting for them. It all just takes too long to explain. A PR consultant might advise organizations discussing energy issues to stick with an easy message: "We are running out of oil," or "We are *not* running out of oil." Take your pick and make your case.

Reality is more complicated.

Fortunately, there is one element of simplicity in all this complexity, at least in terms of communication—and that is *what we must do*: as a global society, we must reduce our dependency on fossil fuels as quickly as possible. It is the only realistic answer both to climate change and our economic vulnerability to declining fossil fuel resource quality and EROEI. This is literally humanity's project of the century, probably the most important in all of history. It is an enormous challenge, but it is not optional. Either we break the addiction, or we suffer the consequences—which would impact not only ourselves, but future generations as well.

Yet, the mistaken notion that new technology can free up all the oil and gas we could ever possibly want stops us in our tracks. Suddenly we are faced with a (false) binary choice: jobs and economic growth on one hand, climate protection on the other.

People need jobs and businesses need growth. If plentiful fossil fuels can provide jobs and growth (we tend to believe they can because they have a track record, and we already have the infrastructure to use those fuels), then can't we somehow find a way to eat our cake, yet have it too? "Let's think about this for a while longer before making any rash decisions," the masses murmur in unison. In this context, "a while" could mean a decade or longer. By that time, it will be far too late to begin a successful energy transition.

The choice is rigged. The promise of *economic* fossil energy abundance is a mirage. Like a thirsty desert castaway, we chase that mirage even though it lures us to our doom. Dazzled by the prospects of a hundred years of cheap natural gas or oil independence, we embrace an energy policy of "all of the above" that is hardly distinguishable from having no energy policy at all. With every passing year the fossil fuel industry consumes a larger portion of global GDP, reducing society's ability to fund an energy transition. And every year the environmental costs of continued fossil fuel reliance compound.

Everything depends upon our recognizing the mirage for what it is, and getting on with the project of the century.

NOTES

INTRODUCTION

1. International Energy Agency, *World Energy Outlook 2000*, www .worldenergyoutlook.org/media/weowebsite/2008-1994/weo2000 .pdf.
2. See, for example, my own book: Richard Heinberg, *The Party's Over: Oil, War and the Fate of Industrial Societies* (Gabriola Island, BC: New Society Publishers, 2003).
3. See, for example, Julian Darley, *High Noon for Natural Gas: The New Energy Crisis* (White River Junction, VT: Chelsea Green Publishing Company, 2004).
4. George Monbiot, "We Were Wrong on Peak Oil. There's Enough to Fry Us All," *Guardian*, July 2, 2012, http://www.guardian.co.uk /commentisfree/2012/jul/02/peak-oil-we-we-wrong.
5. Energy Information Administration, *Annual Energy Outlook 2013 Early Release*, Table 14, (December 5, 2012).

CHAPTER 1

1. Nick Owen, Oliver Inderwildi, and David King, "The Status of Conventional World Oil Reserves—Hype or Cause for Concern?" *Energy Policy* 38, no. 8 (August 2010): 4743–4749, doi: 10.1016/j .enpol.2010.02.026.
2. International Energy Agency, *World Energy Outlook 2008*; see also http://www.postpeakliving.com/files/shared/Hook-GOF_decline _Article.pdf.
3. Matt Mushalik, "Shrinking Crude Oil Exports a Tough Game for Oil Importers," *Crude Oil Peak* (blog), February 4, 2013, http://

crudeoilpeak.info/shrinking-crude-oil-exports-a-tough-game-for
-oil-importers. Mushalik references JODI Oil World Database data
up to November 2012.

4. "Why Is US Oil Consumption Lower? Better Gas Mileage?" *The
Oil Drum* (website), last modified February 6, 2013, http://www
.theoildrum.com/node/9811.

5. "Annual Vehicle-Miles Traveled in the United States and Year
-over-Year Changes, 1971-2012," *The Geography of Transport Systems*
(website), http://people.hofstra.edu/geotrans/eng/ch3en/conc3en
/vehiclemilesusa.html.

6. Jeffrey Brown, "The Export Capacity Index (ECI): A New Metric
For Predicting Future Supplies of Global Net Oil Exports," ASPO
-USA (website), February 18, 2013, http://aspousa.org/2013/02
/commentary-the-export-capacity-index/.

7. Megan C. Guilford et al., "A New Long Term Assessment of Energy
Return on Investment (EROI) for U.S. Oil and Gas Discovery
and Production," *Sustainability* 2011, 3, 1866–1887, doi:10.3390
/su3101866.

8. "White's Law," *Wikipedia*, last modified July 25, 2012, http://en
.wikipedia.org/wiki/White%27s_law.

9. Gail Tverberg, "How Resource Limits Lead to Financial Collapse,"
Our Finite World (blog), March 29, 2013, http://ourfiniteworld
.com/2013/03/29/how-resource-limits-lead-to-financial-collapse/.

10. Robert Gordon, "Is U.S. Economic Growth Over? Faltering
Innovation Confronts the Six Headwinds," The National Bureau
of Economic Research, NBER Working Paper No. 18315, August,
2012, http://www.nber.org/papers/w18315.

11. Gail Tverberg, "How High Oil Prices Lead to Recession," *Our Finite
World* (blog), January 24, 2013, http://ourfiniteworld.com/2013
/01/24/how-high-oil-prices-lead-to-recession/.

CHAPTER 2

1. Peter A. Dea, President & CEO of Cirque Resources, (speaking at the
ASPO-USA annual conference, Denver, CO, October 10, 2009).

2. "Interview with Boone Pickens," *JobVetka* (website), last modified

March 12, 2012, http://jobvetka.blogspot.com/2012/03/interview
-with-boone-pickens.html.

3. "Natural Gas: Fueling America's Future," Chesapeake Energy (website), accessed May 10, 2013, http://www.chk.com/naturalgas /pages/fueling-americas-future.aspx.

4. Stephen Lacey, "After USGS Analysis, EIA Cuts Estimates of Marcellus Shale Gas Reserves by 80%," *ThinkProgress* (blog), August 26, 2011, http://thinkprogress.org/climate/2011/08/26/305467/usgs -marcellus-shale-gas-estimates-overestimated-by-80/?mobile=nc.

5. Leigh Price, "Origins and Characteristics of the Basin-Centered Continuous-Reservoir Unconventional Oil-Resource Base of the Bakken Source System, Williston Basin" (paper presented to the Energy and Environmental Research Center (EERC)), http:// www.undeerc.org/News-Publications/Leigh-Price-Paper/pdf /TextVersion.pdf.

6. Fred F. Meissner and Richard B. Banks, "Computer Simulation of Hydrocarbon Generation, Migration, and Accumulation Under Hydrodynamic Conditions—Examples from the Williston and San Juan Basins, USA" (oral presentation, AAPG International Conference and Exhibition, Bali, Indonesia, October 15–18, 2000), http://www.searchanddiscovery.com/documents/2005/banks/. Jack Flannery and Jeff Kraus, "Integrated Analysis of the Bakken Petroleum System, U.S. Williston Basin" (poster presentation, AAPG Annual Convention, Houston, TX, April 10–12, 2006), http://www .searchanddiscovery.com/documents/2006/06035flannery/.

7. "USGS Releases New Oil and Gas Assessment for Bakken and Three Forks Formations," US Department of the Interior, April 30, 2013, http://www.doi.gov/news/pressreleases/usgs-releases-new-oil-and -gas-assessment-for-bakken-and-three-forks-formations.cfm.

8. Leonardo Maugeri, "Oil: The Next Revolution," Geopolitics of Energy Project, Belfer Center for Science and International Affairs, John F. Kennedy School of Government, Harvard University, June 2012, http://belfercenter.ksg.harvard.edu/files/Oil-%20The%20 Next%20Revolution.pdf.

9. David Strahan, "Oil Glut Forecaster Maugeri Admits Duff Maths," *David Strahan: Energy Writer* (blog), July 30, 2012, http://www. davidstrahan.com/blog/?p=1570.

10. Frederick Kempe, "America's Geopolitical Gusher," *Thinking Global* (blog), *Reuters*, November 26, 2012, http://blogs.reuters.com /thinking-global/2012/11/26 /americas-geopolitical-gusher/.

11. Angel Gonzales, "Making Sense of the U.S. Oil Boom," *Wall Street Journal*, September 13, 2012, http://online.wsj.com/article/SB10000 872396390444301704577631820865343432.html.

CHAPTER 3

1. Loren Steffy, "Shale or Sham? A Skeptic Speaks Out," *Houston Chronicle*, November 12, 2009, http://www.chron.com/business /steffy/article/Shale-or-sham-A-skeptic-speaks-out-1748427.php.

2. Bill Powers, "US Shale Gas Won't Last Ten Years: Bill Powers," interview by Peter Byrne, *The Energy Report* (website), November 8, 2012, http://www.theenergyreport.com/pub/na/14705.

3. J. David Hughes, "Drill, Baby, Drill: Can Unconventional Fuels Usher in a New Era of Energy Abundance?" *Post Carbon Institute* (website), February 2013, http://www.postcarbon.org/drill-baby-drill/.

4. "Drill, Baby, Drill: Can Unconventional Fuels Usher in a New Era of Energy Abundance?" abstract, *Post Carbon Institute* (website), accessed May 10, 2013, http://www.postcarbon.org/drill-baby-drill/.

5. Rune Likvern, "Is Shale Oil Production from Bakken Headed for a Run with 'The Red Queen'?" *The Oil Drum* (blog), September 25, 2012, http://www.theoildrum.com/node/9506.

6. "Number of Producing Gas Wells," US Energy Information Administration (website), accessed April 30, 2013, http://www.eia .gov/dnav/ng/ng_prod_wells_s1_a.htm.

7. Energy Information Administration, *Annual Energy Outlook 2012*.

8. Hughes, "Drill, Baby, Drill," 50.

9. Rafael Sandrea, "Evaluating Production Potential of Mature US Oil, Gas Shale Plays," *Oil & Gas Journal*, December 3, 2012, www .ogj.com/articles/print/vol-110/issue-12/exploration-development /evaluating-production-potential-of-mature-us-oil.html.

10. Rune Likvern, "Is the Typical NDIC Bakken Tight Oil Well a Sales Pitch?," *The Oil Drum* (blog), April 29, 2013, http://www .theoildrum.com/node/9954.

11. Jaci Conrad Pearson, "It Takes Oil Money to Make Oil Money," *Black Hills Pioneer*, September 19, 2012, http://www.bhpioneer.com /local_news/article_7bd871d0-0274-11e2-8011-001a4bcf887a.html.

12. Ashley Eady, "Oil Industry Insiders Divided Over Longevity, Feasibility of Shale Play," *Lubbock Avalanche–Journal*, February 2, 2013, http://lubbockonline .com/business/2013-02-02/oil-industry -insiders-divided-over-longevity-feasability-shale-play?v=135987 3091#.UTDmLBmRrEM.

13. Hughes, "Drill, Baby, Drill," 95.

14. Hughes, "Drill, Baby, Drill," 99.

15. Hughes, "Drill, Baby, Drill," 106.

16. Alison Vekshin, "California's Fracking Bonanza May Fall Short of Promise," *Bloomberg.com*, April 9, 2013, http://www.bloomberg.com /news/2013-04-10/california-s-fracking-bonanza-may-fall-short -of-promise.html.

17. Wael Mahdi, "Saudi Arabia's Shale Plans May Be Slowed by Lack of Water," *Bloomberg.com*, March 12, 2013, http://www.bloomberg .com/news/2013-03-12/saudi-arabia-s-shale-plans-may-be-slowed -by-lack-of-water.html.

18. Jeff Tollefson, "China Slow to Tap Shale-Gas Bonanza," *Nature* 494, no. 7437 (February 20, 2013): 294, doi: 10.1038/49429.

19. Hughes, "Drill, Baby, Drill," 28.

20. US Energy Information Administration, "Annual Energy Outlook Retrospective Review: Evaluation of 2012 and Prior Reference Case Projections," March 2013, http://www.eia.gov/forecasts/aeo /retrospective/pdf/retrospective.pdf.

21. Terry Macalister, "Key Oil Figures Were Distorted by US Pressure, Says Whistleblower," *Guardian*, November 9, 2009, http://www. guardian.co.uk/environment/2009/nov/09/peak-oil-international -energy-agency.

22. Andrew Nikiforuk, "Why Energy Experts Get Things Wrong So Often," *The Tyee* (website), March 20, 2013, http://thetyee.ca /News/2013/03/20/Energy-Experts/.

23. Raymond T. Pierrehumbert, "The Myth of 'Saudi America'," *Slate. com*, February 6, 2013, http://www.slate.com/articles/health_and _science/science/2013/02/u_s_shale_oil_are_we_headed_to_a _new_era_of_oil_abundance.single.html.

24. John Westwood, "Energy Business Prospects," (lecture, SNS 2012, Norwich, England, March 1, 2012). See slide 8 of Westwood's presentation at http://www.slideshare.net/DouglasWestwood/sns2012 -1-mar-2012-jw-slideshare.

CHAPTER 4

1. Felicity Barringer, "Hydrofracking Could Strain Western Water Resources, Study Finds," *New York Times*, May 2, 2013, http://www. nytimes.com/2013/05/02/science/earth/hydrofracking-could -strain-western-water-resources-study-finds.html.

2. Sandra Postel, "As Oil and Gas Drilling Competes for Water, One New Mexico County Says No," *National Geographic*, May 3, 2013, http://newswatch.nationalgeographic.com/2013/05/02/as-oil-and -gas-drilling-competes-for-water-one-new-mexico-county-says-no/.

3. Ian Urbina, "Regulation Lax as Gas Wells' Tainted Water Hits Rivers," *New York Times*, February 26, 2011, http://www .nytimes.com/2011/02/27/us/27gas.html. "Report: Fracking's 'Radioactive Wastewater' Discharged into Drinking Water Supplies," *Environmental Leader* (website), March 1, 2011, http://www .environmentalleader.com/2011/03/01/report-frackings -radioactive-wastewater-discharged-into-drinking-water-supplies/. Abby Zimet, "Fracking Debris Ten Times Too Radioactive for Hazardous Waste Landfill," *Common Dreams* (blog), April 25, 2013, http://www.commondreams.org/further/2013/04/25-2.

4. Ian Urbina, "Regulation Lax as Gas Wells' Tainted Water Hits Rivers," *New York Times*, February 26, 2011, http://www.nytimes. com/2011/02/27/us/27gas.html. Urbina references EPA documents (2011) obtained by the *New York Times*.

5. Sheila M. Olmstead et al., "Shale Gas Development Impacts on Surface Water Quality in Pennsylvania," Proceedings of the National Academy of Sciences, March 11, 2013, doi:10.1073/pnas.1213871110.

6. Brett Walton, "Study: Shale Gas Fracking Taints Rivers in Pennsylvania," *Circle of Blue* (website), March 21, 2013, http://www .circleofblue.org/waternews/2013/world/study-shows-how-shale -gas-development-taints-rivers-in-pennsylvania/.

7. Anthony R. Ingraffea, "Fluid Migration Mechanisms Due to Faulty Well Design and/or Construction: An Overview and Recent Experiences in the Pennsylvania Marcellus Play," *Physicians, Scientists and Engineers for Healthy Energy* (website), October 2012, 8–9. http://www .damascuscitizensforsustainability.org/wp-content/uploads/2012/11 /PSECementFailureCausesRateAnalysisIngraffea.pdf.

8. Maurice B. Dusseault, Malcolm N. Gray, and Pawel A. Nawrocki, "Why Oilwells Leak: Cement Behavior and Long-Term Consequences," (paper presented at the SPE International Oil and Gas Conference and Exhibition, Beijing, China, November 7–10, 2000) www.scribd .com/doc/65704543/Casing-Leaks.

9. Diane Ryder, "Report on Bainbridge Well Problem Released," *News–Herald,* September 11, 2008, http://www.news-herald.com /articles/2008/09/11/news/doc48c8944d2a537622194837.txt.

10. Stephen G. Osborn et al., "Methane Contamination of Drinking Water Accompanying Gas-well Drilling and Hydraulic Fracturing," *Proceedings of the National Academy of Sciences* 108, no. 20 (May 17, 2011), doi: 10.1073.

11. Deborah Solomon and Russell Gold, "EPA Ties Fracking, Pollution," *Wall Street Journal,* December 9, 2011, http://online.wsj.com/article /SB10001424052970203501304577086472373346232.html.

12. Ellen Cantarow, "The Downwinders: Fracking Ourselves to Death in Pennsylvania," *TomDispatch.com* (blog), May 2, 2013, http://www .commondreams.org/view/2013/05/02-3.

13. Joe Spease, Chairman, Kansas Sierra Club Fracking Committee, *Testimony on Risks of Hydraulic Fracturing,* (testimony, Kansas Legislature), January 31, 2012, http://www.kslegislature.org/li_2012 /b2011_12/committees/misc/ctte_h_engy_utls_1_20120131_04 _other.pdf.

14. Peter Lehner, "Fracking's Dark Side Gets Darker: The Problem of Methane Waste," *Natural Resources Defense Council Staff Blog,* October 15, 2012, http://switchboard.nrdc.org/blogs/plehner/frackings_dark _side_gets_darke.html.

15. Jon Hurdle, "US Gas Drilling Boom Stirs Water Worries," *Reuters,* February 25, 2009, http://uk.reuters.com/article/2009/02/25/us -energy-marcellus-idUKTRE51O3L620090225.

16. Philip Doe, "A Must Read Account of Fracking Colorado,"

EcoWatch (website), March 5, 2013, http://ecowatch.com/2013/must-read-fracking-colorado/.

17. Lisa Song, "First Study of Its Kind Detects 44 Hazardous Air Pollutants at Gas Drilling Sites," *InsideClimate News* (blog), December 3, 2012, http://insideclimatenews.org/news/20121203/natural-gas-drilling-air-pollution-fracking-colorado-methane-benzene-endocrine-health-NMHC-epa-toxic-chemicals.

18. In West Virginia and parts of Pennsylvania, most landowners do not own subsurface mineral rights and thus have no say in whether their land is drilled.

19. Ohio Ecological Food and Farm Association, "Fracking and Farmland: What Farmers and Landowners Need to Know About the Risks to Air, Water, and Land," Fracking Factsheet, September 27, 2011, http://oeffa.org/documents/frackingfactsheetv2.pdf.

20. Ibid.

21. Oswald Bamberger, "Impacts of Gas Drilling on Human and Animal Health," *New Solutions* 22, no.1 (2012): 51–77, doi: 10.2190/NS.22.1.e.

22. Judith Kohler, "Report Says Drilling Threatens Colorado Wildlife," *Aspen Times*, January 20, 2010, http://www.aspentimes.com/news/1426301-113/regional-leadstories-regionalivg-leadstoriesivg.

23. Katy Dunlap, Eastern Water Project Director, Trout Unlimited, *Shale Gas Production and Water Resources in the Eastern United States*, (testimony, US Senate Committee on Energy and Natural Resources, Subcommittee on Water and Power), October 20, 2011, http://www.energy.senate.gov/public/index.cfm/files/serve?File_id=1cbe5c49-aa41-4bec-a6b7-992068c59666. Susan Young, "EPA to Regulate Fracking Waste Water," *Nature News Blog*, October 21, 2011, http://blogs.nature.com/news/2011/10/us_senate_keeping_an_eye_on_fr.html.

24. Katie M. Keranen et al., "Potentially Induced Earthquakes in Oklahoma, USA: Links Between Wastewater Injection and the 2011 Mw 5.7 Earthquake Sequence," *Geology* (first published on March 26, 2013), doi: 10.1130/G34045.1. Becky Oskin, "New Mexico Earthquakes Linked to Wastewater Injection," *Live Science* (website), April 23, 2013, http://www.livescience.com/28983-w-mexico-earthquakes-fracking.html.

25. "Loss of Property Values, Difficulty Getting Mortgages and Home Insurance," *Save Colorado From Fracking* (website), http://www .savecoloradofromfracking.org/harm/propertyvalues.html.

26. Jeff Abbas, "Farming, Foraging and Fracking: Our Fight Against the Machine," *Resilience.org*, March 26, 2013, http://www.resilience .org/stories/2013-03-26/farming-foraging-and-fracking-our-fight -against-the-machine. Steve Horn and Trisha Marczak, "Sand Land: Fracking Industry Mining Iowa's Iconic Sand Bluffs in New Form of Mountaintop Removal," *DeSmogblog.com* (blog), April 30, 2013, http://desmogblog.com/2013/04/30/sand-land-fracking-industry -mining-iowa-s-iconic-sand-bluffs-new-mountaintop-removal.

27. Dr. Wayne Feyereisn, "Potential Public Health Risks of Silica Sand Mining and Processing," (presentation hosted by Concerned Citizens for St. Charles, St. Charles, MN), March 7, 2013, http://www .sandpointtimes.com/Potential-Public-Health-Risks-of-Silica-Sand -Mining-and-Processing.htm.

28. US Environmental Protection Agency, *Inventory of US Greenhouse Gas Emissions and Sinks*: 1990–2010, (Washington, DC, April 15, 2012), Energy 3-5, http://www.epa.gov/climatechange/Downloads /ghgemissions/US-GHG-Inventory-2012-Main-Text.pdf. The EPA measured each source of energy by its carbon (C) intensity in comparison to coal.

29. Robert Howarth, Renee Santoro, and Anthony Ingraffea, "Methane and the Greenhouse-Gas Footprint of Natural Gas from Shale Formations," *Climatic Change* 106, no. 4 (June, 2011): 679–690, doi: 10.1007/s10584-011-0061-5.

30. Tom Zeller Jr., "Studies Say Natural Gas Has Its Own Environmental Problems," *New York Times*, April 11, 2011, http://www.nytimes .com/2011/04/12 /business/energy-environment/12gas.html.

31. Jon Entine, "New York Times Reversal: Cornell University Research Undermines Hysteria Contention That Shale Gas Is 'Dirty'," *Forbes*, March 2, 2012, http://www.forbes.com/sites /jonentine/2012/03/02/new-york-times-reversal-cornell-university -research-undermines-hysteria-contention-that-shale-gas-is-dirty/.

32. Lawrence Cathles et al., "A Commentary On 'The Greenhouse-Gas Footprint Of Natural Gas In Shale Formations' by R. W. Howarth,

R. Santoro, and Anthony Ingraffea," *Climatic Change* 113, no. 2 (July 2012): 525–535, doi: 10.1007/s10584-011-0333-0.

33. Ramón A. Alvarez et al, "Greater Focus Needed On Methane Leakage From Natural Gas Infrastructure," *Proceedings of the National Academy of Sciences* 110, no. 13, (March 11, 2013): 4962–4967, doi:10.1073/pnas.1202407109.

34. David Hughes, "Lifecycle Greenhouse Gas Emissions from Shale Gas Compared to Coal: An Analysis of Two Conflicting Studies," Post Carbon Institute, July 2011, 16–18, http://www.postcarbon.org/reports/PCI-Hughes-NETL-Cornell-Comparison.pdf.

35. Jeff Tollefson, "Methane Leaks Erode Green Credentials of Natural Gas," *Nature* 493, no. 7430 (January 2, 2013), http://www.nature.com/news/methane-leaks-erode-green-credentials-of-natural-gas-1.12123#/b1.

36. Ian J. Laurenzi and Gilbert R. Jersey, "Life Cycle Greenhouse Gas Emissions and Freshwater Consumption of Marcellus Shale Gas," *Environmental Science and Technology* 47, no. 9 (April 2, 2013): 4896–4903, doi: 10.1021.

37. "EPA Methane Report Further Divides Fracking Camps," Associated Press, April 28, 2013, http://www.npr.org/templates/story/story.php?storyId=179638846.

38. Tim McDonnell, "Frackers Are Losing $1.5 Billion Yearly to Leaks," *Mother Jones,* April 5, 2013, http://www.motherjones.com/blue-marble/2013/04/frackers-are-losing-15-billion-yearly-leaks.

CHAPTER 5

1. "Major CEOs Feeling the Recession . . . Somewhat," Associated Press, May 1, 2009, http://www.nbcnews.com/id/30501718/#.UZGK6LWkqQs.

2. Christopher Helman, "The Two Sides of Aubrey McClendon, America's Most Reckless Billionaire," *Forbes*, October 5, 2011, http://www.forbes.com/sites/christopherhelman/2011/10/05/aubrey-mcclendon-chesapeake-billionaire-wildcatter-shale/.

3. Christopher Swann and Robert Cyran, "Did Chesapeake miss Enron lessons?" *Reuters*, May 22, 2012, http://uk.reuters.com/article/2012/05/22 /idUKL1E8GMCTB20120522.

4. Christopher Helman, "Here's What The Analyst Who Uncovered Enron Thinks About Chesapeake," *Forbes*, June 4, 2012, http://www .forbes.com/sites/christopherhelman/2012/06/04/enron-chesapeake -olson/.

5. John Shiffman, Anna Driver, and Brian Grow, "Special Report: the Lavish and Leveraged Life of Aubrey McClendon," *Reuters*, June 7, 2012, http://in.reuters.com/article/2012/06/07/us-chesapeake -mcclendon-profile-idINBRE8560IB20120607.

6. Nicholas Sakelaris, "Aubrey McClendon Back in the Game with New Oil/Gas Company," *Dallas Business Journal*, April 19, 2013, http://www .bizjournals.com/dallas/news/2013/04/19/aubrey-mcclendon-back -in-the-game-with.html.

7. "Benefits of Fracking," *EnergyFromShale* (website), accessed May 13, 2013, http://www.energyfromshale.org/fracking-benefits.

8. Matthew DiLallo, "Can Fracking Benefit Your Community Too?" *Motley Fool*, March 26, 2013, http://www.fool.com/investing /general/2013/03/26/can-fracking-benefit-your-community-too .aspx. Alan Bjerga, "Small Towns Find Fracking Brings Boom, Booming Headaches," *Bloomberg*, March 27, 2013, http://www .bloomberg.com/news/2013-03-27/small-towns-find-fracking -brings-boom-booming-headaches.html.

9. Stephen Herzenberg, "Drilling Deeper into Job Claims: The Actual Contribution of Marcellus Shale to Pennsylvania Job Growth," policy brief, Keystone Research Center, June 20, 2011, http:// keystoneresearch.org/sites/keystoneresearch.org/files/Drilling -Deeper-into-Jobs-Claims-6-20-2011_0.pdf.

10. Susan Christopherson, "The Economic Consequences of Marcellus Shale Gas Extraction: Key Issues," Community & Regional Development Initiative, Cornell University, *CaRDI Reports* no. 11 (September·2011), http://www.greenchoices.cornell.edu/downloads /development/marcellus/Marcellus_CaRDI.pdf. See also "Economic Impacts of Fracking," *Save Colorado from Fracking*, http://www .savecoloradofromfracking.org/harm/economic.html.

11. Tara Lohan, "Resource Curse: Why the Economic Boom That Fracking Promises Will Be a Bust For Most People (Hard Times, USA)," *AlterNet*, March 6, 2013, http://www.alternet.org/hard -times-usa/resource-curse-why-economic-boom-fracking

-promises-will-be-bust-most-people-hard. See Susan Christopherson and Ned Rightor, "How Should We Think About the Economic Consequences of Shale Gas Drilling?" http://www.greenchoices .cornell.edu/downloads/development/marcellus/Marcellus_SC _NR.pdf .

12. Deborah Rogers, "Externalities of Shales: Road Damage," *Energy Policy Forum* (blog), April 1, 2013, http://energypolicyforum .org/2013/04/01/externalities-of-shales-road-damage/.

13. "Increased Crime Rates," *Save Colorado from Fracking*, modified October 3, 2005, http://www.savecoloradofromfracking.org/harm /index.html.

14. Deborah Rogers, "Shale and Wall Street: Was the Decline in Natural Gas Prices Orchestrated?" *Energy Policy Forum*, February 2013, http:// shalebubble.org/wall-street/.

15. Brett Shipp, "Landowners Upset Over Unpaid Royalties in the Barnett Shale," *WFAA-TV* (website), updated October 26, 2012, http://www.wfaa.com/news/investigates/Landowners-upset-over-unpaid-royalties-in-the-Barnett-Shale-175868241.html.

16. Andrea Ahles, "DFW Airport Settles Lawsuit with Chesapeake," *Fort Worth Star-Telegram*, September 7, 2012, http://www.star-telegram. com/2012/09/06/4237684/dfw-airport-settles-lawsuit-with.html.

17. Daniel Yergin, Chairman, IHS Cambridge Energy Research Associates, (testimony, Senate Energy and Natural Resources Committee, Senate Energy Committee), October 4, 2011, http:// press.ihs.com/press-release/energy-power/testimony-daniel-yergin-testimony-senate-energy-and-natural-resources-com.

18. Daniel Yergin (testimony submitted for hearings on *America's Energy Security and Innovation, Subcommittee on Energy and Power of the House Energy and Commerce Committee*), February 5, 2013, http://docs .house.gov/meetings/IF/IF03/20130205/100220/HHRG-113 -IF03-Wstate-YerginD-20130205.pdf.

19. "USA: United LNG, Petronet Reach Main Pass Energy Hub Agreement," *LNG World News*, posted April 25, 2013, http://www .lngworldnews.com/usa-united-lng-petronet-reach-main-pass -energy-hub-agreement/.

20. "US Congressmen Favour Export of Natural Gas to India," *Hindu*

Business Line, April 27, 2013, http://www.thehindubusinessline.com/economy/article4660270.ece.

21. The IEA believes that at a natural gas price of $5 per thousand cubic feet, US utilities will begin switching back to coal. http://business.financialpost.com/2013/05/27/iea-says-u-s-gas-prices-of-us5-could-spur-return-to-coal/?__lsa=f9f8-7e1f.

22. Deborah Rogers, "PA Jobs Numbers Poor in Spite of Marcellus Shale," *Energy Policy Forum*, February 4, 2013, http://www.frackcheckwv.net/2013/02/11/low-job-count-in-pa-marcellus-shale-development/.

23. Deborah Rogers, "Shale and Wall Street," *Energy Policy Forum*, February 2013. Data from U.S. Bureau of Labor Statistics, posted May 8, 2012, http://data.bls.gov/timeseries/CES1021100001?data_tool=XGtable.

24. Brad Plumer, "The US Oil and Gas Boom Has Had a Modest Economic Impact—So Far," *Washington Post*, April 23, 2013, http://www.washingtonpost.com/blogs/wonkblog/wp/2013/04/23/the-oil-and-gas-boom-has-had-a-surprisingly-small-impact-on-the-u-s-economy/.

25. Chesapeake Energy Corporation, Q3 2008 Business Update Call Transcript, October 19, 2008, http://seekingalpha.com/article/100644-chesapeake-energy-corporation-q3-2008-business-update-call-transcript.

26. Deborah Rogers, "Shale and Wall Street," http://shalebubble.org/wp-content/uploads/2013/02/SWS-report-FINAL.pdf.

27. Matthieu Auzanneau, "Total Production by the Top Five Oil Majors Has Fallen by a Quarter Since 2004," *The Oil Drum* (blog), April 19, 2013, http://www.theoildrum.com/node/9946.

28. Steve Andrews, "Commentary: Interview with Steven Kopits," *Peak Oil Review* (April 29, 2013): 6–10, http://aspousa.org/wp-content/files/por130429.pdf.

29. "ExxonMobil: 'We're Losing Our Shirts'," *gCaptain* (blog), June 27, 2012, http://gcaptain.com/exxonmobil-were-losing-shirts/.

30. Clifford Krauss and Eric Lipton, "After the Boom in Natural Gas," *New York Times*, October 20, 2012, http://www.nytimes.com/2012/10/21/business/energy-environment/in-a-natural-gas-glut-big-winners-and-losers.html?pagewanted=all&_r=3&. Philip Bump, "Frackers Struggle While Financiers Make Millions. Sounds

Familiar." *Grist* (blog), October 22, 2012, http://grist.org/news/frackers-struggle-while-financiers-make-millions-sounds-familiar/.

31. Deborah Rogers, "Shale and Wall Street," http://shalebubble.org/wp-content/uploads/2013/02/SWS-report-FINAL.pdf.

CHAPTER 6

1. Chris Nelder, "Are Methane Hydrates Really Going to Change Geopolitics?" *Atlantic*, May 2, 2013, http://www.theatlantic.com/technology/archive/2013/05/are-methane-hydrates-really-going-to-change-geopolitics/275275/.

2. Charles C. Mann, "Yes, Unconventional Fossil Fuels Are That Big of a Deal," *Atlantic*, May 7, 2013, http://www.theatlantic.com/technology/archive/2013/05/yes-unconventional-fossil-fuels-are-that-big-of-a-deal/275616/.

3. Roberto Cesare Callarotti, "Energy Return on Energy Invested (EROI) for the Electrical Heating of Methane Hydrate Reservoirs," *Sustainability* 3, no. 11, (November 7, 2011): 2105–2114. doi: 10.3390/su3112105.

4. Cutler J. Cleveland and Peter A. O'Connor, "Energy Return on Investment (EROI) of Oil Shale," *Sustainability* 3, no. 11 (November 22, 2011): 2307–2322, doi: 10.3390/su3112307.

5. Charles Hall, "Unconventional Oil: Tar Sands and Shale Oil—EROI on the Web, Part 3 of 6," *The Oil Drum* (blog), posted by Nate Hagens, April 15, 2008, http://www.theoildrum.com/node/3839.

6. Jacob Chamberlain, "Deeper Than Deepwater: Shell Plans World's Riskiest Offshore Well," *Common Dreams* (website), May 9, 2013, http://www.commondreams.org/headline/2013/05/09-2.

7. Hughes, "Drill, Baby, Drill," 129.

8. Bryan Walsh, "A Rig Accident Off Alaska Shows the Dangers of Extreme Energy," *Time*, January 2, 2013, http://science.time.com/2013/01/02/a-rig-accident-off-alaska-shows-the-dangers-of-extreme-energy/#ixzz2SuGACsh6. Stephanie Joyce, "Shell Tallies Cost of Kulluk Grounding," *Alaska Public Media* (website), February 1, 2013, http://www.alaskapublic.org/2013/02/01/shell-tallies-cost-of-kulluk-grounding/.

9. See, for example, "Improving Efficiency in Upstream Oil Sands Production," ExxonMobil, http://www.exxonmobil.com/Corporate /energy_production_oilsands.aspx. John Kemp, "Column—Bakken Output May Be Boosted by Closer Oil Wells: Kemp," *Reuters*, May 8, 2013, http://in.reuters.com/article/2013/05/08/column-kemp-us -oilwells-idINL6N0DP2LW20130508.

10. Francie Diep, "Solar Panels Now Make More Electricity Than They Use," *Popular Science*, April 3, 2013, http://www.popsci.com/science /article/2013-04/solar-panels-now-make-more-electricity-they-use.

11. Doug Hansen and Charles Hall, eds. "New Studies in EROI (Energy Return on Investment)," special issue, *Sustainability* (2011), http://www.mdpi.com/journal/sustainability/special_issues/New _Studies_EROI.

12. Jessica Lambert et al., "EROI of Global Energy Resources," (State University of New York, College of Environmental Science and Forestry, November 2012), http://www.roboticscaucus.org/ENERGY POLICYCMTEMTGS/Nov2012AGENDA/documents/DFID _Report1_2012_11_04-2.pdf.

13. Charles A. S. Hall, "Editorial: Synthesis to Special Issue on New Studies in EROI (Energy Return on Investment)," *Sustainability* 3, no. 12 (December 14, 2011): 2496–2499, doi:10.3390/su3122496. Andrew McKay has proposed a new unit he calls "Petroleum Production per Unit of Effort," or PPUE, which reflects drilling rates, drilling depths, and cost of production. World PPUE improved between 1980 and 2000 but has declined dramatically since 2000. http://www.resilience.org/stories/2013-05-28/drilling-faster-just -to-stay-still-a-proposal-to-use-production-per-unit-effort-ppue -as-an-indicator-of-peak-oil.

14. Andrew Lees, "In Search of Energy," in *The Gathering Storm*, ed. Patrick Young (Derivatives Vision Publishing, 2010).

15. "Engine Trouble: A Rise in Energy Costs Will Hit Productivity," *Economist*, October 21, 2010, http://www.economist.com/node /17314626?subjectid= 2512631&story_id=17314626.

16. Tim Morgan, "Perfect Storm: Energy, Finance, and the End of Growth," *Tullett Prebon* (blog), January 2013, 77, ftalphaville.ft.com /files/2013/01/Perfect-Storm-LR.pdf.

17. Bryan Sell, David Murphy, and Charles A. S. Hall, "Energy Return

on Energy Invested for Tight Gas Wells in the Appalachian Basin, United States of America," *Sustainability* 3, no. 10 (October 20, 2011), doi: 10.3390/su3101986. Caveats are from private communications with one of the study's authors.

18. Hughes, "Drill, Baby, Drill," 75.
19. For further discussion of this point, citing failures to improve efficiency in tar sands operations, see Andrew Nikiforuk, "Difficult Truths about 'Difficult Oil.'" http://www.resilience.org/stories /2013-05-23/difficult-truths-about-difficult-oil.
20. The EROEI for tight oil production in the Bakken play is under investigation; a report by Egan Waggoner on the subject is in preparation.
21. Morgan, "Perfect Storm," 3.
22. Improvement in EROEI can be inferred from falling prices for new solar and wind installed capacity (private communication with Charles Hall). However, some renewable energy technologies achieve higher EROEI by relying on materials such as rare earth minerals that have an increasing energy cost over time due to depletion of the more accessible deposits. Also, as the best locations for wind turbines, tidal, and geothermal power are utilized, further expansion requires the use of less favorable locations, resulting in lower EROEI.
23. "Renewable Electricity Futures Study," National Renewable Energy Laboratory, last updated May 13, 2013, http://www.nrel.gov /analysis/re_futures/. One early reader of this chapter commented: "You don't necessarily need the same amount of energy to achieve the same functionality post-fossil fuels. For example, in our plug-in vehicles we drive on about one-fifth of the energy used by a typical gas car to achieve the same result of moving people down the road. Plus we make that renewable energy by PV on our own rooftop for one-eighth the cost of gasoline. So you could say we only have one-fifth the energy available to us and paint a negative picture of having 80% less energy available, but we're achieving the same motive result as a fossil fuel powered tool."
24. Benedikt Römer et al., "The Role of Smart Metering and Decentralized Electricity Storage for Smart Grids: The Importance of Positive Externalities," *Energy Policy* 50 (November 2012): 486–495, http:// www.sciencedirect.com/science/article/pii/S0301421512006416. Jan

von Appen, "Time in the Sun: The Challenge of High PV Penetration in the German Electric Grid," *IEEE Power and Energy* 11, no. 2 (March 2013): 55–64, doi: 10.1109/MPE.2012.2234407. For a more optimistic perspective on the potential of microgrids to enable higher levels of renewable energy, see Chris Nelder, "Microgrids: A Utility's Best Friend or Worst Enemy?" http://www.resilience.org/stories/2013-05-24 /microgrids-a-utility-s-best-friend-or-worst-enemy.

25. http://www.iea.org/topics/renewables/.

26. David Manners, "Massive Consolidation in Solar," *Electronics Weekly*, January 14, 2013, http://www.electronicsweekly.com/news/business /massive-consolidation-in-solar-2013-01/.

27. Andrew Herndon, "Biofuel Pioneer Forsakes Renewables to Make Gas-Fed Fuels," *Bloomberg.com*, May 1, 2013, http://www.bloomberg .com/news/2013-04-30/biofuel-pioneer-forsakes-renewables-to -make-gas-fed-fuels.html.

28. Louis Bergeron, "The World Can Be Powered by Alternative Energy, Using Today's Technology, in 20–40 Years, Says Stanford Researcher Mark Z. Jacobson," *Stanford Report*, January 26, 2011, http://news .stanford.edu/news/2011/january/jacobson-world-energy-012611 .html. Amory Lovins, "A 40-year Plan for Energy," TED talk (March 2012) http://www.ted.com/talks/amory_lovins_a_50_year_plan _for_energy.html.

29. Ted Trainer, "Renewable Energy Cannot Sustain a Consumer Society," (Dordrecht, The Netherlands: Springer, 2010). For a moderate and realistic take on the capabilities and limits of renewable energy, see David McKay, *Sustainable Energy—Without the Hot Air* (blog), http://www.withouthotair.com/.

30. Prices could fall absent a full-fledged global recession, if energy efficiency in transport vehicles increases significantly (we are already seeing modest gains) and vehicle miles traveled decrease significantly in regions experiencing very low economic growth.

31. Gail Tverberg, "Low Oil Prices Lead to Economic Peak Oil," *Our Finite World* (blog), April 21, 2013, http://ourfiniteworld .com/2013/04/21/low-oil-prices-lead-to-economic-peak-oil/.

32. Richard Heinberg, *Blackout: Coal, Climate and the Last Energy Crisis* (British Columbia, Canada: New Society Publishers, 2009). Tadeusz Patzek and Gregory Croft, "A Global Coal Production Forecast with

Multi-Hubbert Cycle Analysis," *Energy* 35, no. 8 (August, 2010): 3109–3122, http://www.sciencedirect.com/science/article/pii/S0360544210000617.

33. "The Dream that Failed," *Economist*, March 10, 2012, http://www.economist.com/node/21549936.

34. Gail Tverberg, "How Resource Limits Lead to Financial Collapse," *Our Finite World* (blog), March 29, 2013, http://ourfiniteworld.com/2013/03/29/how-resource-limits-lead-to-financial-collapse/.

35. This, by the way, would not solve serious ecological problems such as resource depletion, topsoil loss, species extinctions, and water scarcity. I'm focusing here only on our energy-economic-climate conundrum.

ABBREVIATIONS

Btu—British thermal unit
EIA—Energy Information Administration
EPA—Environmental Protection Agency
EROEI—energy return on energy invested
EROI—energy return on investment
GDP—gross domestic product
IEA—International Energy Agency
LNG—liquefied natural gas
mb/d—million barrels per day
mcf—thousand cubic feet
NGLs—natural gas liquids
NOAA—National Oceanic and Atmospheric Administration
PR—public relations
tcf—trillion cubic feet
USGS—United States Geological Survey

GLOSSARY

crude oil—As used herein, conventional crude oil not including natural gas liquids, biofuels, or refinery gains.

horizontal well—A well typically started vertically, which is curved to horizontal at depth to follow a particular rock stratum or reservoir.

hydraulic fracturing ("fracking")—The process of inducing fractures in reservoir rocks through the injection of water and other fluids, chemicals, and solids under very high pressure.

multi-stage hydraulic-fracturing—Each individual hydraulic fracturing treatment is a "stage" localized to a portion of the well. There may be as many as 30 individual hydraulic fracturing stages in some wells.

oil shale—Organic-rich rock that contains kerogen, a precursor of oil; not to be confused with shale oil. Depending on organic content, it can sometimes be burned directly with a calorific value equivalent to a very low-grade coal. Can be "cooked" in situ at high temperatures for several years to produce oil or can be retorted in surface operations to produce petroleum liquids.

petroleum liquids (also, "liquids")—All petroleum-like liquids used as liquid fuels, including crude oil, lease condensates, natural gas liquids, refinery gains, and biofuels.

play—A prospective area for the production of oil, gas, or both. Usually a relatively small contiguous geographic area focused on an individual reservoir.

reserve—A deposit of oil, gas, or coal that can be recovered profitably within existing economic conditions using existing technologies. Has legal implications in terms of company valuations for the Securities and Exchange Commission.

shale gas—Gas contained in shale with very low permeabilities in the micro- to nano-darcy range. Typically produced using horizontal wells with multi-stage hydraulic fracture treatments.

shale oil—See **tight oil**.

stripper well—An oil or gas well that is nearing the end of its economically useful life. In the United States, a "stripper" gas well is defined by the Interstate Oil and Gas Compact Commission as one that produces 60,000 cubic feet (1,700 m3) or less of gas per day at its maximum flow rate. Oil wells are generally classified as stripper wells when they produce 10 barrels per day or less for any 12-month period.

tight oil—Also referred to as shale oil. Oil contained in shale and associated clastic and carbonate rocks with very low permeabilities in the micro- to nano-darcy range. Typically produced using horizontal wells with multi-stage hydraulic fracture treatments.

type decline curve—The average production declines for all wells in a given area or play from the first month on production. For shale plays in this study, the type decline curves considered the average of the first four to five years of production.

undiscovered technically recoverable resource—Resources inferred to exist using probabilistic methods extrapolated from available exploration data and discovery histories. Usually designated with confidence levels. For example, P90 indicates a 90% chance of having a least the stated resource volume whereas a P10 estimate has only a 10% chance.

FURTHER ACKNOWLEDGMENTS

The Merry Band of Editors sponsored the production of this book, reviewed early drafts, and provided critical feedback. Many thanks to each of these dedicated folks for their support and their input!

Sherine Adeli
Kristin Adkins
Chatral A'dzé
Franky Aelbrecht
Will Alexander
Mark Allcock
Nick Allen
Per O. Andersson
Tommy Andreasen
Bob Armantrout
Victoria Armigo
Martin Astrand
Matt Austin
Robin Baena
Bill Ballou
Marilyn Bardet
Brad Bardwell
Vidura Barrios
Kathleen Basman
Nancy Bell
Craig Benjamin
Edwin Benson
Michael Benson
Mark Berger
Desmond Berghofer
Howard B. Bernstein

John Berton
Andy Bevan
Karam Bhogal
Glenn Bier
David Binar
Cindy Blackshear
Jeff Blackshear
Mark Bloore
Christof Bojanowski
Darrel Bostow
Rob Branch-Dasch
Moshe Braner
Peter Brezny
Carolyn Bridge
Tod Edwin Brilliant
Paul Bristow
Michael Brock
Ray Broggini
Richard Brook
Cal Broomhead
Karen Brown
Marilyn Brown
Michael Brownlee
Hank Brummer
Andy Buckingham
Grant George Buffett

LeeAnne Burton
Ruth Busch
Roberto Campanaro
Anna M. Campbell
Bill Campbell
Frank Campbell
Dolly Carlisle
Ann Carranza
Nicholas Carter
Kate Case
Tim Castle
Gary Charbonneau
Leslie Christian
Lars Christiansen
David Christopher
Anthony Christy
Clifton P. Chute
Peter Clare
Barbara Clark
Doug Close
Gary Coates
Michael J. Coe
Craig K. Comstock
Debbie Cook
Sonia Corbett
Robin Curtis
Tim Cuthbertson
Carolina Dahlberg
Bob Daniel
Mariquita de Boissiere
John de Jardin
Deborah Deal-Blackwell
Earl Dean
Dwain Deets
E. D. Dennis
Nancy Deren
Silvia Di Blasio

Jed Diamond
Leo DiDomenico
Angel Dobrow
Ollie Downward
Daniel du Toit
Kendall Dunnigan
John Duvall
Paul Eagle
Chris Eames
Janet M. Eaton
C. Peter Eckrich
Brett Eisenlohr
Emerging Technology Corporation
Colin Endean
Kay Engler
Thomas Everth
Piero Falotti
Thomas Fellows
Tom Ferris
Ed Fields
Dave Finnigan
Linda Fiolich
Robert Fischer
Gloria Flora
W. R. Flynn
Peter Follett
Peter Foster-Bunch
James Freund
Isaias Galvez
Richard Geray
Jon Gething
Paul F. Getty
George Girod
Jenny Goldie
Jeff Goldman
Robert Goldschmidt
Patricia Goldsmith

Daphne Golliher
Fred G. Gregory
Steve Hackenberg
Linda Hagan
Linton Hale
Caroline Hancock
Phil Hardy
Karey Harrison
Tian Harter
M'Lynn Hartwell
Kirsten Hasberg
Guy Haslam
Matthew Havens
Paul Hawley
John T. Heinen
Edward Hejtmanek
Toby Hemenway
Douglas Hendren
Warren Hendricks
Brad Herrlinger
Sara Hess
Yashi Hoffman
John Hoffmann
Scott Honn
Richard Hookway
John Howe
Bob Hutchinson
George Hylkema
Don Hynes
Judith Iam
David Iandiorio
Fred E. Irwin
M. Jackson
Veronica Jacobi
Franke James
Piere Jason
Peter Jensen

Andy Johnson
Joyelle Jolie
Maurice Jones
James Kalin
Robert Kaulfuss
Donald S. Kelly
Carol Kennedy
Deirdre Kent
Cooper J. Kessel
Nicole Kindred
Joseph Kinner
Danny Kirkeby
Erv Klaas
William M. Klassen
Herb Kline
Nancy Klock
Mary Kobe
Janet Kobren
Mike Kotschenreuther
Lisi Krall
Bruce LaCour
Maureen Lafreniere
Cajup Lalinca
John Lamb
Vane Lashua
Kim Latham
Cameron Leckie
Daniel Lerch
Vicki Lipski
Peter Loomans
Nicolas Louchet
Oberg Lyle
Robert Magill
William Maiden
Alex Malcolm
Greg Mann
Anna Manzo

Lynne Mao

Hazel Marchant

Patrick Marchman

Trisha Marlow

Brian Marsden

Luke Evans Massman-Johnson

Edward S. Matalka

Keith Maw

Thomas Maxwell

Jackson McCarty

Peter McClelland

Fred McColly

Bruce McDonald

Donald McKim

Larry Menkes

Bernie Meyer

Steve Meyer

John Miglietta

Adam Miller

Chris Mills

Andrew Milne

William Minatre

Jackie Minchew

Jim Horne Minter

Scott Mittelsteadt

Pierre Montminy

David Moorhouse

Daniel Morinigo-Sotelo

Guy Morse-Brown

Tom Mundahl

Kim Mundell

Michael Mussotter

Aaron Naparstek

Eva Naylor

Adam Nealis

Peter Neils

Gerardus Neve

Alfred Nye, Jr.

Michael O'Hara

Christina Olsen

Mikael Olsson

Clarice Ondrack

R. Orman

Ann Pacey

Miroslaw Pacocha

Erin Pammenter

Nathan J. Parkin

Rauli Partanen

Janet Patterson

Roger Peck

Luca Ferrari Pedraglio

Dave Petersen

James Peterson

Mark Petry

Bonnie Petty

Laurel Phoenix

Daniel Pickles

Stanislas Pique

Benjamin Pittenger

Susan Porter

Ted Pounder

James Raymond

Mat Redsell

Bethany Reece

Justin Ritchie

Mark Robinowitz

Caleb Rockenbaugh

Ann Rogers

Paula Rohrbaugh

Maria Rotunda

Denise Rushing

Rachel Sachs

Roland Saher

Djordje Samardzija

Lee Samelson
Gary Sanders
Brian Sanderson
Karl Schmid
Ward Schmidt
Clifford Dean Scholz
David Schonbrunn
Linda Schweitzer
Jack Scott
Paul Sebert
Sean Seefried
Darren Shupe
Luiz Mauricio de Miranda e Silva
Ernie Simpson
Deborah Sims
Peter Sims
John Sleeman
Dave Smalley
William Dean Smith
Brent Smith
Norton Smith
Jonathan Smolens
Shane Derek Snell
Richard Stauffacher
Gerrit Stegehuis
Jan Steinman
Steve Stevens
Greg Swan
Charley Sweet
Martha Taranto
Brian Thompson
Ibo Thorbas
Jody Tishmack
Ron Tjerandsen
Rick Toyne
Roy C. Treadway
Gerald Tremblay

Benjamin Trister
Jeffry Troeger
D. Bruce Turton
Robert Van Every
Ashwani Vasishth
Paul Vidovich
George Vye
Celia Fulton Walden
Ian Warder
N.G. Ware
David Warrender
Ruth White
Donovan C. Wilkin
Andrew Willner
Donna Wilson
Nancy Lee Wood
Marion Yaglinski
Monowaruz Zaman
Sander Zegveld
Miriam Zolin